Real-Resumes For Safety & Quality Assurance Jobs

...including real resumes used to change careers
and resumes used to gain federal employment

Anne McKinney, Editor

PREP PUBLISHING

FAYETTEVILLE, NC

PREP Publishing
1110 ½ Hay Street
Fayetteville, NC 28305
(910) 483-6611

Library of Congress Cataloging-in-Publication Data

Real-resumes for safety & quality assurance jobs--: including real resumes used to change careers and transfer skills to other industries / Anne McKinney, editor.
p. cm. -- (Real-resumes series)
ISBN 1-885288-45-X
1. Resumes (Employment)--Handbooks, manuals, etc. 2. Safety consultants--Vocational guidance. 3. Quality assurance--Vocational guidance. I. Title: Real-resumes for safety and quality assurance jobs. II. McKinney, Anne, 1948- III. Series.

HF5383.R39595 2005
650.14'2--dc22 2005043107

Printed in the United States of America

PREP Publishing

Business and Career Series:

RESUMES AND COVER LETTERS THAT HAVE WORKED, Revised Edition
RESUMES AND COVER LETTERS THAT HAVE WORKED FOR MILITARY PROFESSIONALS
GOVERNMENT JOB APPLICATIONS AND FEDERAL RESUMES
COVER LETTERS THAT BLOW DOORS OPEN
LETTERS FOR SPECIAL SITUATIONS
RESUMES AND COVER LETTERS FOR MANAGERS
REAL-RESUMES FOR COMPUTER JOBS
REAL-RESUMES FOR MEDICAL JOBS
REAL-RESUMES FOR FINANCIAL JOBS
REAL-RESUMES FOR TEACHERS
REAL-RESUMES FOR STUDENTS
REAL-RESUMES FOR CAREER CHANGERS
REAL-RESUMES FOR SALES
REAL ESSAYS FOR COLLEGE & GRADUATE SCHOOL
REAL-RESUMES FOR AVIATION & TRAVEL JOBS
REAL-RESUMES FOR POLICE, LAW ENFORCEMENT & SECURITY JOBS
REAL-RESUMES FOR SOCIAL WORK & COUNSELING JOBS
REAL-RESUMES FOR CONSTRUCTION JOBS
REAL-RESUMES FOR MANUFACTURING JOBS
REAL-RESUMES FOR RESTAURANT, FOOD SERVICE & HOTEL JOBS
REAL-RESUMES FOR MEDIA, NEWSPAPER, BROADCASTING & PUBLIC AFFAIRS JOBS
REAL-RESUMES FOR RETAILING, MODELING, FASHION & BEAUTY JOBS
REAL-RESUMES FOR HUMAN RESOURCES & PERSONNEL JOBS
REAL-RESUMES FOR NURSING JOBS
REAL-RESUMES FOR AUTO INDUSTRY JOBS
REAL RESUMIX & OTHER RESUMES FOR FEDERAL GOVERNMENT JOBS
REAL KSAS--KNOWLEDGE, SKILLS & ABILITIES--FOR GOVERNMENT JOBS
REAL BUSINESS PLANS & MARKETING TOOLS
REAL-RESUMES FOR ADMINISTRATIVE SUPPORT, OFFICE & SECRETARIAL JOBS
REAL-RESUMES FOR FIREFIGHTING JOBS
REAL-RESUMES FOR JOBS IN NONPROFIT ORGANIZATIONS
REAL-RESUMES FOR SPORTS INDUSTRY JOBS
REAL-RESUMES FOR LEGAL & PARALEGAL JOBS
REAL-RESUMES FOR ENGINEERING JOBS
REAL-RESUMES FOR U.S. POSTAL SERVICE JOBS
REAL-RESUMES FOR CUSTOMER SERVICE JOBS
REAL-RESUMES FOR SAFETY & QUALITY ASSURANCE JOBS

Judeo-Christian Ethics Series:

SECOND TIME AROUND
BACK IN TIME
WHAT THE BIBLE SAYS ABOUT...Words that can lead to success and happiness
A GENTLE BREEZE FROM GOSSAMER WINGS
BIBLE STORIES FROM THE OLD TESTAMENT

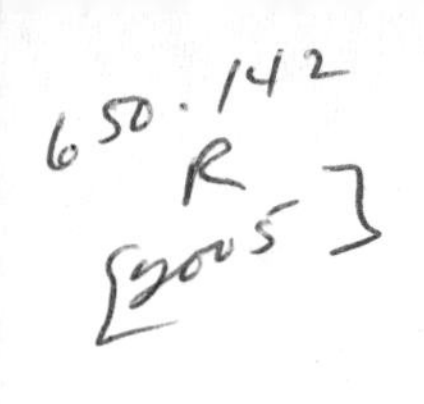

Contents

Real-Resumes For Safety & Quality Assurance Jobs

Anne McKinney, Editor

A WORD FROM THE EDITOR: ABOUT THE REAL-RESUMES SERIES

Welcome to the Real-Resumes Series. The Real-Resumes Series is a series of books which have been developed based on the experiences of real job hunters and which target specialized fields or types of resumes. As the editor of the series, I have carefully selected resumes and cover letters (with names and other key data disguised, of course) which have been used successfully in real job hunts. That's what we mean by "Real-Resumes." What you see in this book are *real* resumes and cover letters which helped real people get ahead in their careers.

We hope the superior samples will help you manage your current job campaign and your career so that you will find work aligned to your career interests.

The Real-Resumes Series is based on the work of the country's oldest resume-preparation company known as PREP Resumes. If you would like a free information packet describing the company's resume preparation services, call 910-483-6611 or write to PREP at 1110½ Hay Street, Fayetteville, NC 28305. If you have a job hunting experience you would like to share with our staff at the Real-Resumes Series, please contact us at preppub@aol.com or visit our website at www.prep-pub.com.

The resumes and cover letters in this book are designed to be of most value to people already in a job hunt or contemplating a career change. If we could give you one word of advice about your career, here's what we would say: Manage your career and don't stumble from job to job in an incoherent pattern. Try to find work that interests you, and then identify prosperous industries which need work performed of the type you want to do. Learn early in your working life that a great resume and cover letter can blow doors open for you and help you maximize your salary.

Introduction: The Art of Changing Jobs... and Finding New Careers

As the editor of this book, I would like to give you some tips on how to make the best use of the information you will find here. Because you are considering a career change, you already understand the concept of managing your career for maximum enjoyment and self-fulfillment. The purpose of this book is to provide expert tools and advice so that you *can* manage your career. Inside these pages you will find resumes and cover letters that will help you find not just a job but the type of work you want to do.

Overview of the Book

Every resume and cover letter in this book actually worked. And most of the resumes and cover letters have common features: most are one-page, most are in the chronological format, and most resumes are accompanied by a companion cover letter. In this section you will find helpful advice about job hunting. Step One begins with a discussion of why employers prefer the one-page, chronological resume. In Step Two you are introduced to the direct approach and to the proper format for a cover letter. In Step Three you learn the 14 main reasons why job hunters are not offered the jobs they want, and you learn the six key areas employers focus on when they interview you. Step Four gives nuts-and-bolts advice on how to handle the interview, send a follow-up letter after an interview, and negotiate your salary.

The cover letter plays such a critical role in a career change. You will learn from the experts how to format your cover letters and you will see suggested language to use in particular career-change situations. It has been said that "A picture is worth a thousand words" and, for that reason, you will see numerous examples of effective cover letters used by real individuals to change fields, functions, and industries.

The most important part of the book is the Real-Resumes section. Some of the individuals whose resumes and cover letters you see spent a lengthy career in an industry they loved. Then there are resumes and cover letters of people who wanted a change but who probably wanted to remain in their industry. Many of you will be especially interested by the resumes and cover letters of individuals who knew they definitely wanted a career change but had no idea what they wanted to do next. Other resumes and cover letters show individuals who knew they wanted to change fields and had a pretty good idea of what they wanted to do next.

Whatever your field, and whatever your circumstances, you'll find resumes and cover letters that will "show you the ropes" in terms of successfully changing jobs and switching careers.

Before you proceed further, think about why you picked up this book.

- Are you dissatisfied with the type of work you are now doing?
- Would you like to change careers, change companies, or change industries?
- Are you satisfied with your industry but not with your niche or function within it?
- Do you want to transfer your skills to a new product or service?
- Even if you have excelled in your field, have you "had enough"? Would you like the stimulation of a new challenge?
- Are you aware of the importance of a great cover letter but unsure of how to write one?
- Are you preparing to launch a second career after retirement?
- Have you been downsized, or do you anticipate becoming a victim of downsizing?
- Do you need expert advice on how to plan and implement a job campaign that will open the maximum number of doors?
- Do you want to make sure you handle an interview to your maximum advantage?

- Would you like to master the techniques of negotiating salary and benefits?
- Do you want to learn the secrets and shortcuts of professional resume writers?

Using the Direct Approach

As you consider the possibility of a job hunt or career change, you need to be aware that most people end up having at least three distinctly different careers in their working lifetimes, and often those careers are different from each other. Yet people usually stumble through each job campaign, unsure of what they should be doing. Whether you find yourself voluntarily or unexpectedly in a job hunt, the direct approach is the job hunting strategy most likely to yield a full-time permanent job. The direct approach is an active, take-the-initiative style of job hunting in which you choose your next employer rather than relying on responding to ads, using employment agencies, or depending on other methods of finding jobs. You will learn how to use the direct approach in this book, and you will see that an effective cover letter is a critical ingredient in using the direct approach.

The "direct approach" is the style of job hunting most likely to yield the maximum number of job interviews.

Lack of Industry Experience Not a Major Barrier to Entering New Field

"Lack of experience" is often the last reason people are not offered jobs, according to the companies who do the hiring. If you are changing careers, you will be glad to learn that experienced professionals often are selling "potential" rather than experience in a job hunt. Companies look for personal qualities that they know tend to be present in their most effective professionals, such as communication skills, initiative, persistence, organizational and time management skills, and creativity. Frequently companies are trying to discover "personality type," "talent," "ability," "aptitude," and "potential" rather than seeking actual hands-on experience, so your resume should be designed to aggressively present your accomplishments. Attitude, enthusiasm, personality, and a track record of achievements in any type of work are the primary "indicators of success" which employers are seeking, and you will see numerous examples in this book of resumes written in an all-purpose fashion so that the professional can approach various industries and companies.

Using references in a skillful fashion in your job hunt will inspire confidence in prospective employers and help you "close the sale" after interviews.

The Art of Using References in a Job Hunt

You probably already know that you need to provide references during a job hunt, but you may not be sure of how and when to use references for maximum advantage. You can use references very creatively during a job hunt to call attention to your strengths and make yourself "stand out." Your references will rarely get you a job, no matter how impressive the names, but the way you use references can boost the employer's confidence in you and lead to a job offer in the least time.

You should ask from three to five people, including people who have supervised you, if you can use them as a reference during your job hunt. You may not be able to ask your current boss since your job hunt is probably confidential.

A common question in resume preparation is: "Do I need to put my references on my resume?" No, you don't. Even if you create a references page at the same time you prepare your resume, you don't need to mail, e-mail, or fax your references page with the resume and cover letter. Usually the potential employer is not interested in references until he meets you, so the earliest you need to have references ready is at the first interview. Obviously there are exceptions to this standard rule of thumb; sometimes an ad will ask you to send references with your first response. Wait until the employer requests references before providing them.

An excellent attention-getting technique is to take to the first interview not just a page of references (giving names, addresses, and telephone numbers) but an actual letter of reference written by someone who knows you well and who preferably has supervised or employed you. A professional way to close the first interview is to thank the interviewer, shake his or her hand, and then say you'd like to give him or her a copy of a letter of reference from a previous employer. Hopefully you already made a good impression during the interview, but you'll "close the sale" in a dynamic fashion if you leave a letter praising you and your accomplishments. For that reason, it's a good idea to ask supervisors during your final weeks in a job if they will provide you with a written letter of recommendation which you can use in future job hunts. Most employers will oblige, and you will have a letter that has a useful "shelf life" of many years. Such a letter often gives the prospective employer enough confidence in his opinion of you that he may forego checking out other references and decide to offer you the job on the spot or in the next few days.

With regard to references, it's best to provide the names and addresses of people who have supervised you or observed you in a work situation.

Whom should you ask to serve as references? References should be people who have known or supervised you in a professional, academic, or work situation. References with big titles, like school superintendent or congressman, are fine, but remind busy people when you get to the interview stage that they may be contacted soon. Make sure the busy official recognizes your name and has instant positive recall of you! If you're asked to provide references on a formal company application, you can simply transcribe names from your references list. In summary, follow this rule in using references: If you've got them, flaunt them! If you've obtained well-written letters of reference, make sure you find a polite way to push those references under the nose of the interviewer so he or she can hear someone other than you describing your strengths. Your references probably won't ever get you a job, but glowing letters of reference can give you credibility and visibility that can make you stand out among candidates with similar credentials and potential!

The approach taken by this book is to (1) help you master the proven best techniques of conducting a job hunt and (2) show you how to stand out in a job hunt through your resume, cover letter, interviewing skills, as well as the way in which you present your references and follow up on interviews. Now, the best way to "get in the mood" for writing your own resume and cover letter is to select samples from the Table of Contents that interest you and then read them. A great resume is a "photograph," usually on one page, of an individual. If you wish to seek professional advice in preparing your resume, you may contact one of the professional writers at Professional Resume & Employment Publishing (PREP) for a brief free consultation by calling 1-910-483-6611.

STEP ONE: Planning Your Career Change and Assembling the Tools

Part One: Some Advice About Your Job Hunt

What if you don't know what you want to do?
Your job hunt will be more comfortable if you can figure out what type of work you want to do. But you are not alone if you have no idea what you want to do next! You may have knowledge and skills in certain areas but want to get into another type of work. What *The Wall Street Journal* has discovered in its research on careers is that most of us end up having at least three distinctly different careers in our working lives; it seems that, even if we really like a particular kind of activity, twenty years of doing it is enough for most of us and we want to move on to something else!

Figure out what interests you and you will hold the key to a successful job hunt and working career. (And be prepared for your interests to change over time!)

That's why we strongly believe that you need to spend some time figuring out ***what interests you*** rather than taking an inventory of the skills you have. You may have skills that you simply don't want to use, but if you can build your career on the things that interest you, you will be more likely to be happy and satisfied in your job. Realize, too, that interests can change over time; the activities that interest you now may not be the ones that interested you years ago. For example, some professionals may decide that they've had enough of retail sales and want a job selling another product or service, even though they have earned a reputation for being an excellent retail manager. We strongly believe that interests rather than skills should be the determining factor in deciding what types of jobs you want to apply for and what directions you explore in your job hunt. Obviously one cannot be a lawyer without a law degree or a secretary without secretarial skills; but a professional can embark on a next career as a financial consultant, property manager, plant manager, production supervisor, retail manager, or other occupation if he/she has a strong interest in that type of work and can provide a resume that clearly demonstrates past excellent performance in *any* field and *potential* to excel in another field. As you will see later in this book, "lack of exact experience" is the last reason why people are turned down for the jobs they apply for.

"Lack of exact experience" is the last reason people are turned down for the jobs for which they apply.

How can you have a resume prepared if you don't know what you want to do?
You may be wondering how you can have a resume prepared if you don't know what you want to do next. The approach to resume writing which PREP, the country's oldest resume-preparation company, has used successfully for many years is to develop an "all-purpose" resume that translates your skills, experience, and accomplishments into language employers can understand. What most people need in a job hunt is a versatile resume that will allow them to apply for numerous types of jobs. For example, you may want to apply for a job in pharmaceutical sales but you may also want to have a resume that will be versatile enough for you to apply for jobs in the construction, financial services, or automotive industries.

Based on more than 20 years of serving job hunters, we at PREP have found that your best approach to job hunting is **an all-purpose resume** and **specific cover letters tailored to specific fields** rather than using the approach of trying to create different resumes for every job. If you are remaining in your field, you may not even need more than one "all-purpose" cover letter, although the cover letter rather than the resume is the place to communicate your interest in a narrow or specific field. An all-purpose resume and cover letter that translate your experience and accomplishments into plain English are the tools that will maximize the number of doors which open for you while permitting you to "fish" in the widest range of job areas.

Your resume will provide the script for your job interview.

When you get down to it, your resume has a simple job to do: Its purpose is to blow as many doors open as possible and to make as many people as possible want to meet you. So a well-written resume that really "sells" you is a key that will create opportunities for you in a job hunt.

This statistic explains why: The typical newspaper advertisement for a job opening receives more than 245 replies. And normally only 10 or 12 will be invited to an interview.

But here's another purpose of the resume: it provides the "script" the employer uses when he interviews you. If your resume has been written in such a way that your strengths and achievements are revealed, that's what you'll end up talking about at the job interview. Since the resume will govern what you get asked about at your interviews, you can't overestimate the importance of making sure your resume makes you look and sound as good as you are.

Your resume is the "script" for your job interviews. Make sure you put on your resume what you want to talk about or be asked about at the job interview.

So what is a "good" resume?

Very literally, your resume should motivate the person reading it to dial the phone number or e-mail the screen name you have put on the resume. When you are relocating, you should put a local phone number on your resume if your physical address is several states away; employers are more likely to dial a local telephone number than a long-distance number when they're looking for potential employees.

If you have a resume already, look at it objectively. Is it a limp, colorless "laundry list" of your job titles and duties? Or does it "paint a picture" of your skills, abilities, and accomplishments in a way that would make someone want to meet you? Can people understand what you're saying? If you are attempting to change fields or industries, can potential employers see that your skills and knowledge are transferable to other environments? For example, have you described accomplishments which reveal your problem-solving abilities or communication skills?

The one-page resume in chronological format is the format preferred by most employers.

How long should your resume be?

One page, maybe two. Usually only people in the academic community have a resume (which they usually call a *curriculum vitae*) longer than one or two pages. Remember that your resume is almost always accompanied by a cover letter, and a potential employer does not want to read more than two or three pages about a total stranger in order to decide if he wants to meet that person! Besides, don't forget that the more you tell someone about yourself, the more opportunity you are providing for the employer to screen you out at the "first-cut" stage. A resume should be concise and exciting and designed to make the reader want to meet you in person!

Should resumes be functional or chronological?

Employers almost always prefer a chronological resume; in other words, an employer will find a resume easier to read if it is immediately apparent what your current or most recent job is, what you did before that, and so forth, in reverse chronological order. A resume that goes back in detail for the last ten years of employment will generally satisfy the employer's curiosity about your background. Employment more than ten years old can be shown even more briefly in an "Other Experience" section at the end of your "Experience" section. Remember that your intention is not to tell everything you've done but to "hit the high points" and especially impress the employer with what you learned, contributed, or accomplished in each job you describe.

STEP TWO: Using Your Resume and Cover Letter

Once you get your resume, what do you do with it?
You will be using your resume to answer ads, as a tool to use in talking with friends and relatives about your job search, and, most importantly, in using the "direct approach" described in this book.

When you mail your resume, always send a "cover letter."
A "cover letter," sometimes called a "resume letter" or "letter of interest," is a letter that accompanies and introduces your resume. Your cover letter is a way of personalizing the resume by sending it to the specific person you think you might want to work for at each company. Your cover letter should contain a few highlights from your resume—just enough to make someone want to meet you. Cover letters should always be typed or word processed on a computer—never handwritten.

Never mail or fax your resume without a cover letter.

1. Learn the art of answering ads.
There is an "art," part of which can be learned, in using your "bestselling" resume to reply to advertisements.

Sometimes an exciting job lurks behind a boring ad that someone dictated in a hurry, so reply to any ad that interests you. Don't worry that you aren't "25 years old with an MBA" like the ad asks for. Employers will always make compromises in their requirements if they think you're the "best fit" overall.

What about ads that ask for "salary requirements?"
What if the ad you're answering asks for "salary requirements?" The first rule is to avoid committing yourself in writing at that point to a specific salary. You don't want to "lock yourself in."

There are two ways to handle the ad that asks for "salary requirements."
First, you can ignore that part of the ad and accompany your resume with a cover letter that focuses on "selling" you, your abilities, and even some of your philosophy about work or your field. You may include a sentence in your cover letter like this: "I can provide excellent personal and professional references at your request, and I would be delighted to share the private details of my salary history with you in person."

What if the ad asks for your "salary requirements?"

Second, if you feel you must give some kind of number, just state a range in your cover letter that includes your medical, dental, other benefits, and expected bonuses. You might state, for example, "My current compensation, including benefits and bonuses, is in the range of $30,000-$40,000."

Analyze the ad and "tailor" yourself to it.
When you're replying to ads, a finely tailored cover letter is an important tool in getting your resume noticed and read. On the next page is a cover letter which has been "tailored to fit" a specific ad. Notice the "art" used by PREP writers of analyzing the ad's main requirements and then writing the letter so that the person's background, work habits, and interests seem "tailor-made" to the company's needs. Use this cover letter as a model when you prepare your own reply to ads.

Date

Exact Name of Person
Title or Position
Name of Company
Address
City, State, Zip

Dear Exact Name of Person: (or Dear Sir or Madam if answering a blind ad.)

I would like to make you aware of my interest in a Quality Assurance position within a corporation which can utilize my strong executive skills. I offer a proven ability to apply my technical expertise in resourceful ways that improve the bottom line while strengthening customer satisfaction.

Employers are trying to identify the individual who wants the job they are filling. Don't be afraid to express your enthusiasm in the cover letter!

As you will see from my resume, I have worked for the past 17 years for Best Products Company, where I have been promoted to increasing levels of responsibility. In my current job, I manage a 19-person QA department in a plant which employs 425 people and manufactures personal care products totaling $250 million. While managing a departmental budget of nearly $1 million, I have transitioned the plant from a regular production assembly line operation into a team-managed operation in which teams of employees are responsible for individual products. This has shifted QA from a "police" role to a consulting and monitoring role. I have also developed, implemented and managed a Cost-of-Quality Program which achieved 2004 cost savings of $150,000 by identifying and eliminating unnecessary processes.

In my previous job at Quality Assurance Manager at the company's Massachusetts plant, I developed and implemented a Quality Demerit System which the company now uses corporate-wide. The Quality Demerit System transformed the four National Industry manufacturing plants from a quality level of 67% defect-free product to the consumer in 2003, to 98.2% defect-free in 2004. The targeted goal for 2005 is 99.0% defect-free product.

I offer extensive expertise related to blow molding and injection molding. Both in my current job and in my job in Massachusetts, I managed Quality Assurance related to blow molding and injection molding. In the Massachusetts plant, we achieved a 10% improvement in lots accepted when using Mil. Std. 105E to determine acceptable quality levels (AQLs).

You would find me in person to be a congenial individual who prides myself on my ability to establish and maintain effective working relationships. I can provide outstanding personal and professional references at the appropriate time. I hope you will contact me to suggest a time when we might meet to discuss your needs as well as my skills, experience, and qualifications.

Sincerely,

Raymond Peterson

2. Talk to friends and relatives.
Don't be shy about telling your friends and relatives the kind of job you're looking for. Looking for the job you want involves using your network of contacts, so tell people what you're looking for. They may be able to make introductions and help set up interviews.

About 25% of all interviews are set up through "who you know," so don't ignore this approach.

3. Finally, and most importantly, use the "direct approach."
More than 50% of all job interviews are set up by the "direct approach." That means you actually mail, e-mail, or fax a resume and a cover letter to a company you think might be interesting to work for.

The "direct approach" is a strategy in which you choose your next employer.

To whom do you write?
In general, you should write directly to the *exact name* of the person who would be hiring you: say, the vice-president of marketing or data processing. If you're in doubt about to whom to address the letter, address it to the president by name and he or she will make sure it gets forwarded to the right person within the company who has hiring authority in your area.

How do you find the names of potential employers?
You're not alone if you feel that the biggest problem in your job search is finding the right names at the companies you want to contact. But you can usually figure out the names of companies you want to approach by deciding first if your job hunt is primarily geography-driven or industry-driven.

In a **geography-driven job hunt,** you could select a list of, say, 50 companies you want to contact **by location** from the lists that the U.S. Chambers of Commerce publish yearly of their "major area employers." There are hundreds of local Chambers of Commerce across America, and most of them will have an 800 number which you can find through 1-800-555-1212. If you and your family think Atlanta, Dallas, Ft. Lauderdale, and Virginia Beach might be nice places to live, for example, you could contact the Chamber of Commerce in those cities and ask how you can obtain a copy of their list of major employers. Your nearest library will have the book which lists the addresses of all chambers.

In an **industry-driven job hunt,** and if you are willing to relocate, you will be identifying the companies which you find most attractive in the industry in which you want to work. When you select a list of companies to contact **by industry,** you can find the right person to write and the address of firms by industrial category in *Standard and Poor's, Moody's,* and other excellent books in public libraries. Many Web sites also provide contact information.

Many people feel it's a good investment to actually call the company to either find out or double-check the name of the person to whom they want to send a resume and cover letter. It's important to do as much as you feasibly can to assure that the letter gets to the right person in the company.

On-line research will be the best way for many people to locate organizations to which they wish to send their resume. It is outside the scope of this book to teach Internet research skills, but librarians are often useful in this area.

What's the correct way to follow up on a resume you send?
There is a polite way to be aggressively interested in a company during your job hunt. It is ideal to end the cover letter accompanying your resume by saying, "I hope you'll welcome my call next week when I try to arrange a brief meeting at your convenience to discuss your current and future needs and how I might serve them." Keep it low key, and just ask for a "brief meeting," not an interview. Employers want people who show a determined interest in working with them, so don't be shy about following up on the resume and cover letter you've mailed.

It pays to be aware of the 14 most common pitfalls for job hunters.

STEP THREE: Preparing for Interviews

But a resume and cover letter by themselves can't get you the job you want. You need to "prep" yourself before the interview. Step Three in your job campaign is "Preparing for Interviews." First, let's look at interviewing from the hiring organization's point of view.

What are the biggest "turnoffs" for potential employers?
One of the ways to help yourself perform well at an interview is to look at the main reasons why organizations *don't* hire the people they interview, according to those who do the interviewing.

Notice that "lack of appropriate background" (or lack of experience) is the *last* reason for not being offered the job.

The 14 Most Common Reasons Job Hunters Are Not Offered Jobs *(according to the companies who do the interviewing and hiring):*

1. Low level of accomplishment
2. Poor attitude, lack of self-confidence
3. Lack of goals/objectives
4. Lack of enthusiasm
5. Lack of interest in the company's business
6. Inability to sell or express yourself
7. Unrealistic salary demands
8. Poor appearance
9. Lack of maturity, no leadership potential
10. Lack of extracurricular activities
11. Lack of preparation for the interview, no knowledge about company
12. Objecting to travel
13. Excessive interest in security and benefits
14. Inappropriate background

Department of Labor studies have proven that smart, "prepared" job hunters can increase their beginning salary while getting a job in *half* the time it normally takes. (4½ months is the average national length of a job search.) Here, from PREP, are some questions that can prepare you to find a job faster.

Are you in the "right" frame of mind?
It seems unfair that we have to look for a job just when we're lowest in morale. Don't worry *too* much if you're nervous before interviews. You're supposed to be a little nervous, especially if the job means a lot to you. But the best way to kill unnecessary

fears about job hunting is through 1) making sure you have a great resume and 2) preparing yourself for the interview. Here are three main areas you need to think about before each interview.

Do you know what the company does?

Don't walk into an interview giving the impression that, "If this is Tuesday, this must be General Motors."

Research the company before you go to interviews.

Find out before the interview what the company's main product or service is. Where is the company heading? Is it in a "growth" or declining industry? (Answers to these questions may influence whether or not you want to work there!)

Information about what the company does is in annual reports, in newspaper and magazine articles, and on the Internet. If you're not yet skilled at Internet research, just visit your nearest library and ask the reference librarian to guide you to printed materials on the company.

Do you know what you want to do for the company?

Before the interview, try to decide how you see yourself fitting into the company. Remember, "lack of exact background" the company wants is usually the last reason people are not offered jobs.

Understand before you go to each interview that the burden will be on you to "sell" the interviewer on why you're the best person for the job and the company.

How will you answer the critical interview questions?

Anticipate the questions you will be asked at the interview, and prepare your responses in advance.

Put yourself in the interviewer's position and think about the questions you're most likely to be asked. Here are some of the most commonly asked interview questions:

Q: *"What are your greatest strengths?"*

A: Don't say you've never thought about it! Go into an interview knowing the three main impressions you want to leave about yourself, such as "I'm hard-working, loyal, and an imaginative cost-cutter."

Q: *"What are your greatest weaknesses?"*

A: Don't confess that you're lazy or have trouble meeting deadlines! Confessing that you tend to be a "workaholic" or "tend to be a perfectionist and sometimes get frustrated when others don't share my high standards" will make your prospective employer see a "weakness" that he likes. Name a weakness that your interviewer will perceive as a strength.

Q: *"What are your long-range goals?"*

A: If you're interviewing with Microsoft, don't say you want to work for IBM in five years! Say your long-range goal is to be *with* the company, contributing to its goals and success.

Q: *"What motivates you to do your best work?"*

A: Don't get dollar signs in your eyes here! "A challenge" is not a bad answer, but it's a little cliched. Saying something like "troubleshooting" or "solving a tough problem" is more interesting and specific. Give an example if you can.

Q: "What do you know about this organization?"

A: Don't say you never heard of it until they asked you to the interview! Name an interesting, positive thing you learned about the company recently from your research. Remember, company executives can sometimes feel rather "maternal" about the company they serve. Don't get onto a negative area of the company if you can think of positive facts you can bring up. Of course, if you learned in your research that the company's sales seem to be taking a nose-dive, or that the company president is being prosecuted for taking bribes, you might politely ask your interviewer to tell you something that could help you better understand what you've been reading. Those are the kinds of company facts that can help you determine whether or not you want to work there.

Go to an interview prepared to tell the company why it should hire you.

Q: "Why should I hire you?"

A: "I'm unemployed and available" is the wrong answer here! Get back to your strengths and say that you believe the organization could benefit by a loyal, hard-working cost-cutter like yourself.

In conclusion, you should decide in advance, before you go to the interview, how you will answer each of these commonly asked questions. Have some practice interviews with a friend to role-play and build your confidence.

STEP FOUR: Handling the Interview and Negotiating Salary

Now you're ready for Step Four: actually handling the interview successfully and effectively. Remember, the purpose of an interview is to get a job offer.

A smile at an interview makes the employer perceive of you as intelligent!

Eight "do's" for the interview

According to leading U.S. companies, there are eight key areas in interviewing success. You can fail at an interview if you mishandle just one area.

1. **Do wear appropriate clothes.**
You can never go wrong by wearing a suit to an interview.

2. **Do be well groomed.**
Don't overlook the obvious things like having clean hair, clothes, and fingernails for the interview.

3. **Do give a firm handshake.**
You'll have to shake hands twice in most interviews: first, before you sit down, and second, when you leave the interview. Limp handshakes turn most people off.

4. **Do smile and show a sense of humor.**
Interviewers are looking for people who would be nice to work with, so don't be so somber that you don't smile. In fact, research shows that people who smile at interviews are perceived as more intelligent. So, smile!

5. **Do be enthusiastic.**
Employers say they are "turned off" by lifeless, unenthusiastic job hunters who show no special interest in that company. The best way to show some enthusiasm for the employer's operation is to find out about the business beforehand.

6. **Do show you are flexible and adaptable.**
An employer is looking for someone who can contribute to his organization in a flexible, adaptable way. No matter what skills and training you have, employers know every new employee must go through initiation and training on the company's turf. Certainly show pride in your past accomplishments in a specific, factual way ("I saved my last employer $50.00 a week by a new cost-cutting measure I developed"). But don't come across as though there's nothing about the job you couldn't easily handle.

7. **Do ask intelligent questions about the employer's business.**
An employer is hiring someone because of certain business needs. Show interest in those needs. Asking questions to get a better idea of the employer's needs will help you "stand out" from other candidates interviewing for the job.

8. **Do "take charge" when the interviewer "falls down" on the job.**
Go into every interview knowing the three or four points about yourself you want the interviewer to remember. And be prepared to take an active part in leading the discussion if the interviewer's "canned approach" does not permit you to display your "strong suit." You can't always depend on the interviewer's asking you the "right" questions so you can stress your strengths and accomplishments.

Employers are seeking people with good attitudes whom they can train and coach to do things their way.

An important "don't": Don't ask questions about salary or benefits at the first interview.
Employers don't take warmly to people who look at their organization as just a place to satisfy salary and benefit needs. Don't risk making a negative impression by appearing greedy or self-serving. The place to discuss salary and benefits is normally at the second interview, and the employer will bring it up. Then you can ask questions without appearing excessively interested in what the organization can do for you.

Now...negotiating your salary
Even if an ad requests that you communicate your "salary requirement" or "salary history," you should avoid providing those numbers in your initial cover letter. You can usually say something like this: "I would be delighted to discuss the private details of my salary history with you in person."

Once you're at the interview, you must avoid even appearing *interested* in salary before you are offered the job. Make sure you've "sold" yourself before talking salary. First show you're the "best fit" for the employer and then you'll be in a stronger position from which to negotiate salary. **Never** bring up the subject of salary yourself. Employers say there's no way you can avoid looking greedy if you bring up the issue of salary and benefits before the company has identified you as its "best fit."

Don't appear excessively interested in salary and benefits at the interview.

Interviewers sometimes throw out a salary figure at the first interview to see if you'll accept it. You may not want to commit yourself if you think you will be able to negotiate a better deal later on. Get back to finding out more about the job. This lets the interviewer know you're interested primarily in the job and not the salary.

When the organization brings up salary, it may say something like this: "Well, Mary, we think you'd make a good candidate for this job. What kind of salary are we talking about?" You may not want to name a number here, either. Give the ball back to the interviewer. Act as though you hadn't given the subject of salary much thought and respond something like this: "Ah, Mr. Jones, I wonder if you'd be kind enough to tell me what salary you had in mind when you advertised the job?" Or ... "What is the range you have in mind?"

Don't worry, if the interviewer names a figure that you think is too low, you can say so without turning down the job or locking yourself into a rigid position. The point here is to negotiate for yourself as well as you can. You might reply to a number named by the interviewer that you think is low by saying something like this: "Well, Mr. Lee, the job interests me very much, and I think I'd certainly enjoy working with you. But, frankly, I was thinking of something a little higher than that." That leaves the ball in your interviewer's court again, and you haven't turned down the job either, in case it turns out that the interviewer can't increase the offer and you still want the job.

Salary negotiation can be tricky.

Last, send a follow-up letter.
Mail, e-mail, or fax a letter right after the interview telling your interviewer you enjoyed the meeting and are certain (if you are) that you are the "best fit" for the job. The people interviewing you will probably have an attitude described as either "professionally loyal" to their companies, or "maternal and proprietary" if the interviewer also owns the company. In either case, they are looking for people who want to work for *that* company in particular. The follow-up letter you send might be just the deciding factor in your favor if the employer is trying to choose between you and someone else. You will see an example of a follow-up letter on page 16.

A follow-up letter can help the employer choose between you and another qualified candidate.

A cover letter is an essential part of a job hunt or career change.
Many people are aware of the importance of having a great resume, but most people in a job hunt don't realize just how important a cover letter can be. The purpose of the cover letter, sometimes called a **"letter of interest,"** is to introduce your resume to prospective employers. The cover letter is often the critical ingredient in a job hunt because the cover letter allows you to say a lot of things that just don't "fit" on the resume. For example, you can emphasize your commitment to a new field and stress your related talents. The cover letter also gives you a chance to stress outstanding character and personal values. On the next two pages you will see examples of very effective cover letters.

A cover letter is an essential part of a career change.

Please do not attempt to implement a career change without a cover letter. A cover letter is the first impression of you, and you can influence the way an employer views you by the language and style of your letter.

Special help for those in career change
We want to emphasize again that, especially in a career change, the cover letter is very important and can help you "build a bridge" to a new career. A creative and appealing cover letter can begin the process of encouraging the potential employer to imagine you in an industry other than the one in which you have worked.

As a special help to those in career change, there are resumes and cover letters included in this book which show valuable techniques and tips you should use when changing fields or industries. The resumes and cover letters of career changers are identified in the table of contents as "Career Change" and you will see the "Career Change" label on cover letters in Part Two where the individuals are changing careers.

***Looking Closer:* The ANATOMY OF A COVER LETTER**

Addressing the Cover Letter: Get the exact name of the person to whom you are writing. This makes your approach personal.

First Paragraph: This explains why you are writing.

Second Paragraph: You have a chance to talk about whatever you feel is your most distinguishing feature.

Third Paragraph: You bring up your next most distinguishing qualities and try to sell yourself.

Fourth Paragraph: Here you have another opportunity to reveal qualities or achievements which will impress your future employer.

Final Paragraph: She asks the employer to contact her. Make sure your reader knows what the "next step" is.

Alternate Final Paragraph: It's more aggressive (but not too aggressive) to let the employer know that you will be calling him or her. Don't be afraid to be persistent. Employers are looking for people who know what they want to do.

Date

Exact Name of Person
Title or Position
Name of Company
Address
City, State, Zip

Dear Exact Name of Person: (or Dear Sir or Madam if answering a blind ad.)

With the enclosed resume, I would like to make you aware of my background as an experienced, self-motivated, and educated industrial and manufacturing engineer with excellent communication and organizational skills and a background in manufacturing, quality assurance, and safety.

In my most recent position as a Senior Manufacturing Engineer at Westfield, I was responsible for the security and quality assurance of more than $9 million worth of equipment, and developed and managed a $1.5 million budget for the maintenance department. Through proper process and materials selection, I reduced hazardous waste generated by the plant by 99%. I also oversaw the removal and proper disposal of all hazardous materials from the paint line and paint booth after that operation ceased.

In previous positions with Westfield, I was responsible for training employees on environmental, health, and safety issues and personally addressing any problems in these areas. Through my initiative, the facility drastically reduced its generation of hazardous waste, which resulted in a downgrading of our Hazardous Waste Generator Status to Conditionally Exempt. I also developed and implemented a Safety First Program that was credited with reducing accidents and improving employees' safety knowledge.

My innovative ideas were displayed when, as a Senior Manufacturing Engineer, I designed a wiring harness for one of our products, resulting in a $100,000 per year reduction in the manufacturing costs for that item. As Management Systems Supervisor, I managed three departments simultaneously, including the product transfer team that moved nearly $10 million worth of production materials to Chicago from a facility in another state in order to support the addition of a new product to our production line and the expansion of our operation.

If you can use a highly skilled quality assurance professional with strong communication and problem-solving skills as well as a background which includes hazardous materials handling and safety, I hope you will contact me soon. I can assure you in advance that I have an outstanding reputation and could rapidly become a valuable addition to your company.

Sincerely,

Christine T. Abraham

Date

Exact Name of Person
Title or Position
Name of Company
Address
City, State, Zip

This accomplished professional is responding to an advertisement. She analyzed the job vacancy opening very closely and she has made sure that she has tailored her letter of interest to the areas mentioned in the vacancy announcement.

Dear Exact Name of Person: (or Dear Sir or Madam if answering a blind ad.)

I would appreciate an opportunity to talk with you soon about how I could contribute to your organization through my exceptionally strong problem-solving skills as well as my experience related to safety and quality assurance.

Expertise related to safety and production
As you will see from my resume, I have excelled in jobs which required unusual resourcefulness, creative strategic thinking, and prudent decision making. In my current job directing quality assurance, I have continuously unplugged bottlenecks and solved stubborn problems in supply and logistics. With a reputation as an astute thinker and outstanding writer, I authored and published a 100-page "maintenance award package" which was named "best in the corporation" and is now considered the model for all divisions.

Strong background in quality control
While managing maintenance and production, I have learned to work in an environment in which there is "no room for error" so concepts like quality control and total quality management are truly "second nature" to me. Even in an earlier job as Chief of Administration for a 6,000-person organization, I became known for my strong quality control orientation as I led that 41-year-old organization to achieve an historical "first": it achieved top scores on three consecutive inspections of all areas of operation.

Skills in human resources administration
I have also excelled in top human resources jobs. In one job I became the leader of a newly consolidated headquarters operation supporting more than 600 people, and I rapidly implemented many novel ideas which boosted morale and productivity. On another occasion, I took over a marginal administrative operation and quickly turned it into a highly motivated operation.

You would find me to be a congenial and dynamic professional who is known for selfless dedication to duty. I can provide outstanding personal and professional references upon your request. I hope you will welcome my call soon when I try to arrange a brief meeting at your convenience to discuss your needs and how I might serve them. Thank you in advance for your time.

Sincerely yours,

Jenna Shortt

Date

Exact Name of Person
Title or Position
Name of Company
Address (number and street)
Address (city, state, and zip)

Follow-up Letter

A great follow-up letter can motivate the employer to make the job offer, and the salary offer may be influenced by the style and tone of your follow-up letter, too!

Dear Exact Name:

I am writing to express my appreciation for the time you spent with me on December 9, and I want to let you know that I am sincerely interested in the position of Safety Director which we discussed.

I feel confident that I could skillfully interact with your staff, and I would cheerfully relocate to Tennessee, as we discussed.

As you described to me what you are looking for in the person who fills this position, I had a sense of "déjà vu" because my current employer was in a similar position when I went to work for Kempert Manufacturing. The general manager needed someone to come in and be his "right arm" and take on an increasing amount of his management responsibilities so that he could be freed up to do other things. I have played a key role in the growth and success of the organization's new Quality Assurance Division, and my supervisor has come to depend on my sound advice as much as well as my proven ability to "cut through" huge volumes of work efficiently and accurately. Since this is one of the busiest times of the year for my employer, I feel that I could not leave during that time. I could certainly make myself available by mid-January.

It would be a pleasure to work for your organization, and I am confident that I could contribute significantly through my strong qualities of loyalty, reliability, and trustworthiness. I am confident that I could quickly learn your style and procedures, and I would welcome being trained to do things your way.

Yours sincerely,

Jacob Evangelisto

PART TWO:
Real-Resumes for Safety & Quality Assurance Jobs

In this section, you will find resumes and cover letters of professionals seeking employment, or already employed, in safety and quality assurance fields. How do these individuals differ from other job hunters? Why should there be a book dedicated to people seeking jobs in the safety and quality assurance fields? Based on more than 20 years of experience in working with job hunters, this editor is convinced that resumes and cover letters which "speak the lingo" of the field you wish to enter will communicate more effectively than language which is not industry-specific. This book is designed to help people (1) who are seeking to prepare their own resumes and (2) who wish to use as models "real" resumes of individuals who have successfully launched careers in safety or quality assurance or advanced in those career paths. You will see a wide range of experience levels reflected in the resumes in this book. Some of the resumes and cover letters were used by individuals seeking to enter the field; others were used successfully by senior professionals to advance in the field.

Newcomers to an industry sometimes have advantages over more experienced professionals. In a job hunt, junior professionals can have an advantage over their more experienced counterparts. Prospective employers often view the less experienced workers as "more trainable" and "more coachable" than their seniors. This means that the mature professional who has already excelled in a first career can, with credibility, "change careers" and transfer skills to other industries.

Newcomers to the field may have disadvantages compared to their seniors. Almost by definition, the inexperienced professional—the young person who has recently entered the job market, or the individual who has recently received respected certifications—is less tested and less experienced than senior managers, so the resume and cover letter of the inexperienced professional may often have to "sell" his or her potential to do something he or she has never done before. Lack of experience in the field she wants to enter can be a stumbling block to the junior employee, but remember that many employers believe that someone who has excelled in anything—academics, for example—can excel in many other fields.

Some advice to inexperienced professionals...
If senior professionals could give junior professionals a piece of advice about careers, here's what they would say: Manage your career and don't stumble from job to job in an incoherent pattern. Try to find work that interests you, and then identify prosperous industries which need work performed of the type you want to do. Learn early in your working life that a great resume and cover letter can blow doors open for you and help you maximize your salary.

Industrial manufacturing

Date

Exact Name of Person
Title or Position
Name of Company
Address
City, State, Zip

ACTING QUALITY ASSURANCE MANAGER
with experience as a senior facilities engineer

Dear Exact Name of Person: (or Dear Sir or Madam if answering a blind ad.)

I would appreciate an opportunity to talk with you soon about how I could contribute to your organization through my background of accomplishments and experience in the management of maintenance, construction, and renovation projects.

As you will see from my enclosed resume, I offer a strong history of reducing costs, bringing projects in on time, and handling the complexities of large-scale domestic and off-shore construction and renovation projects. In my most recent position as Senior Facilities Engineer for Mandell Products in San Diego, I oversaw all phases of physical plant expansion and renovation projects for this 1,400-employee manufacturing plant.

In a previous job as the Manager for Facilities Engineering for Fruit of the Loom in San Diego, I provided expertise during a period of major growth and expansion for this consumer goods manufacturer with 65 sites. I managed construction projects throughout the Pacific, and many of my ideas were incorporated into the company's safety and quality assurance guidelines. Prior to my promotion to that position, as a Senior Facilities Engineer I managed a $3.5 million asbestos abatement program, developed and implemented fire protection programs, and made significant contributions to the company's environmental programs and hazardous waste disposal protocols. One of the projects I coordinated was a waste incineration and steam generation start-up for the city of Scottsdale, AZ.

If you can use a positive, results-oriented manager who enjoys challenges and meets deadlines with precision and enthusiasm, I hope you will contact me soon so that we might discuss your needs. I can assure you in advance that I have an excellent reputation and would quickly become a valuable asset to your company.

Sincerely,

Leonard Robinson

LEONARD ROBINSON

1110½ Hay Street, Fayetteville, NC 28305 • preppub@aol.com • (910) 483-6611

OBJECTIVE

To offer a strong background of distinguished accomplishments in the areas of equipment and facilities construction, maintenance, and renovation to an organization that can benefit from my management experience in project engineering and maintenance.

EDUCATION

B.S., Safety & Quality Engineering, University of San Diego, San Diego, CA, 1996.

EXPERIENCE

Have built a reputation as a hard-charging and innovative management professional:

ACTING QUALITY ASSURANCE MANAGER. Mandell Products, San Diego, CA (2002-present). After two years as the manager of 17 electrical and mechanical technicians completing plant maintenance and capital renovations, was promoted to oversee safety and quality assurance in the process of making improvements to the physical plant, including expansion and renovation projects.

- Emphasize safety and provide weekly safety briefings as I administer on-going renovations to an 800,000 square-foot non-union plant with 1,400 employees producing automotive air and oil filters for both original equipment and aftermarket customers.
- Apply my expertise with EPA regulations, fire protection, and safety while reducing property insurance costs and improving employee safety through renovations to the fire and emergency alarm systems.
- Negotiated fees and made spot purchases which reduced expenses for utilities.
- Assisted in the completion and start up of a 500,000-square-foot distribution center.
- Since 2005, when the company expanded its quality assurance program, have served as Acting Quality Assurance Manager. As Maintenance Manager from 2002-04, managed the environmental departments and organized the maintenance department.
- Develop long-range plans for plant HVAC, lighting, and roofing installations.

MANAGER FOR FACILITIES ENGINEERING. Fruit of the Loom, San Diego, CA (1988-02). Was promoted to manage three engineers and oversee the construction and renovation of facilities for this rapidly growing and expanding organization after approximately eight years as Senior Facilities Engineer.

- Made vital contributions which allowed this consumer goods manufacturer to expand to 65 plants; as manager for construction projects in the Pacific Islands, oversaw a $10 million operational budget for the company in 2000. Developed innovative ideas which resulted in a $240,000 cost reduction for property insurance for the 65 facilities.
- Supervised engineers involved in construction, renovation, environmental, and maintenance projects in both domestic and off-shore facilities.
- As Senior Facilities Engineer from 1988-96, provided oversight for capital projects related to plant and equipment improvements including developing and implementing fire protection programs, production improvements, EPA compliance, and expansions.
- Managed a $3.5 million asbestos abatement program for domestic and offshore facilities.
- Reduced plant maintenance budgets by $200,000; developed maintenance management and training programs. Coordinated a waste incineration and steam generation installation and start up with the city of Scottsdale,, AZ.

Highlights of other experience: Excelled in positions which included Director of Maintenance, Engineering, and Construction; Plant Industrial Engineer and Division Material Handling Engineer; Facilities Planning and Projects Engineer; and Field Engineer.

PERSONAL

Am available for relocation according to employer needs. Enjoy traveling and becoming familiar with other cultures. Respond to challenge with determination to excel.

Automobile manufacturing

Date

Exact Name of Person
Title or Position
Name of Company
Address
City, State, Zip

AREA MANAGER & TOTAL QUALITY CONTROL COORDINATOR

Dear Exact Name of Person: (or Dear Sir or Madam if answering a blind ad.)

With the enclosed resume, I would like to make you aware of my interest in exploring employment opportunities with your organization.

As you will see from my resume, I offer an extensive background in safety and quality assurance in the automotive industry. Immediately after graduating from college, I joined the Goodyear Tire Company, where I learned a great deal about safety and quality assurance in an environment that was oriented towards achieving high goals for productivity while exhibiting a commitment to safety "first, last, and always."

In 1998 I was recruited by my current employer, General Motors, to join its management team. After spending a year as a Management Trainee, I was promoted to Assembly Manager. In that capacity, I managed 32 technicians involved in assembling vital automotive components, and I transformed the department's safety record from unsatisfactory to outstanding through new safety initiatives which I implemented. Because of my success in that position, I was selected for promotion to my current job as Area Manager and Total Quality Management Coordinator. I have been widely credited with creating "a quality culture" with this major automobile manufacturing facility, and I have thoroughly enjoyed playing a role in improving the plant's safety results. Through my creativity and resourceful problem-solving style, I have continuously discovered new ways to reduce waste, improve safety, and optimize quality production.

Although I am held in high regard in my current position and can provide outstanding references at the appropriate time, I am selectively exploring opportunities in other companies which possess a vigorous commitment to safety and quality assurance.

I hope you will call or write me soon to suggest a time convenient for us to meet to discuss your current and future needs. Thank you in advance for your time.

Sincerely yours,

Thomas McIntyre

Alternate last paragraph:

I hope you will welcome my call soon to arrange a brief meeting when we might meet to discuss your needs and goals and how my background might serve them. I can provide outstanding references at the appropriate time.

THOMAS McINTYRE

1110½ Hay Street, Fayetteville, NC 28305 • preppub@aol.com • (910) 483-6611

OBJECTIVE

To contribute to the growth and financial success of an organization that can make use of my background in safety program management, Total Quality Management, and production supervision along with my reputation as an insightful problem solver.

EDUCATION

Bachelor of Science degree in Industrial Engineering Technology, University of Mobile, AL, 1996.

Have completed numerous professional development programs sponsored by local colleges and by General Motors including the following:

Supervisory Interpersonal Skills	Managing for Productivity
Conducting Work Group Meetings	Road Map to Problem Solving
Y.R.I.Q. Instructor Training	Plant Trainer Certification Course

EXPERIENCE

Have become a valuable member of the General Motors Company, Mobile, AL, and have excelled in the following positions:

AREA MANAGER & TOTAL QUALITY MANAGEMENT COORDINATOR. (2005-present). As Area Manager for a job shop operation that is an integral part of new automobile production, am supervising 38 skilled technicians while continuously monitoring production quality, production costs, safety, quality assurance, and numerous factors related to supervision including absenteeism, attitude, and the overall esprit de corps of area associates.

- On my own initiative, assumed responsibilities as a Total Quality Management Coordinator; develop lesson plans related to safety and conduct classes for up to 125 plant personnel at least once a week. Have been credited with playing a key role in implementing "a quality culture" within the plant.
- Am respected for my management style which relies heavily on actively communicating with associates in order to remain sensitive to new opportunities for safety improvements.
- Was the catalyst in resurrecting the department's defunct Ergonomics Task Force; now act as the leader of this task force which meets weekly to identify new opportunities for promoting safety awareness, reducing accidents, and eliminating work hazards.
- Discovered new ways to reduce waste, improve safety, and optimize quality production.

ASSEMBLY MANAGER. (1999-05). As Assembly Manager in charge of supervising 32 technicians, was cited as a major contributing factor to the department's setting new records for production as well as lowest scrap levels.

- Excelled in supervising and motivating associates in a high-production environment; instilled in associates new appreciation for the company's high standards not only in production but also in safety and Total Quality Management.

MANAGER TRAINEE. (1998-99). Rotated to numerous assignments within this huge plant while gaining an overview of the organizational structure and getting to know personnel in all departments. Became aware of the company's emphasis on quality assurance, and learned that everyone in the organization has to make quality his/her personal responsibility.

Other experience: QUALITY ASSURANCE INSPECTOR. Goodyear Tires, Mobile, AL (1996-97). Was recruited for this position prior to college graduation. Was cited as the reason for a significant increase in quality output of ten operators wile overseeing the production of tires for automobiles, including wheeled and track vehicles.

PERSONAL

Would welcome international assignments. Am very knowledgeable of Central American cultures. Fluently speak, read, and write German. Outstanding references on request.

Aircraft manufacturing

Date

Exact Name of Person
Title or Position
Exact Name of Company
Address
City, State, Zip

AVIATION SAFETY CHIEF

Dear Exact Name of Person: (or Dear Sir or Madam if answering a blind ad.)

With the enclosed resume, I would like to express my interest in exploring employment opportunities with your organization.

As you will see from my resume, I have excelled in the aviation safety and quality assurance field while working for one of the best known manufacturers of rotary wing aircraft, Sikorsky, Inc. In my current position as Aviation Safety Chief, I am overseeing plant safety as teams of employees work together to manufacture state-of-the-art Apache helicopters, many of which will be purchased by the U.S. Army for combat operations. Currently working at the company's Alabama plant, I previously excelled as a Team Leader and Quality Assurance Inspector at the corporate headquarters in Connecticut. In that capacity I maintained flawless accountability of multimillion-dollar aircraft parts as I performed strict quality assurance inspections.

Prior to my employment with Sikorsky, I worked for Sikorsky's main customer: the U.S. Army. As a Helicopter Mechanic, I became proficient in keeping helicopters operational in the combat environment of Iraq. I was the recipient of numerous medals and awards recognizing my leadership and creativity as a manager, and I was credited with saving time and money through procedures I established for maintenance and troubleshooting.

Since embarking on my employment with Sikorsky, I have used my spare time in the evenings and on weekends to earn my Bachelor's degree in Industrial Safety. As graduation approaches, I am selectively approaching companies that are leaders in safety and quality assurance.

I hope you will welcome my call soon to arrange a brief meeting to discuss your current and future needs and how I might serve them. Thank you in advance for your time.

Sincerely,

Tristen Hancock

Alternate last paragraph:

I hope you will call or write me soon to suggest a time convenient for us to meet and discuss your current and future needs and how I might serve them. Thank you in advance for your time.

TRISTEN HANCOCK

1110½ Hay Street, Fayetteville, NC 28305 • preppub@aol.com • (910) 483-6611

OBJECTIVE

To contribute to an organization that can a skilled, knowledgeable aviation mechanic with the ability to motivate and supervise and effectively manage time and resources.

EDUCATION

College: Completed 1½ years of study, **Engineering**, Casper College, Casper, WY. In my spare time, am pursuing completion of a B.S. in Industrial Safety, University of Alabama, Tuscaloosa campus; degree anticipated in 2006. Current GPA is 3.8.

Technical Training: Completed Sikorsky's training which included the 15-week Attack Helicopter Repair Course as well as other programs in lifesaving.

Leadership Training: Completed the Army's **Primary Leadership Development Course** which emphasizes leadership, communication, resource management, and technical skills.

AIRCRAFT & TECHNICAL EXPERTISE

Through training and experience, have developed extensive knowledge which includes:

Aircraft: AH-64A Apache helicopter and UH-60 Blackhawk helicopter

Equipment and systems: Use Aviation Ground Power Unit (AGPU), Engine Flush Aviation Equipment, and work on GE 701-C engines (Apache helicopters)

EXPERIENCE

Advanced ahead of my peers to supervisory roles. Sikorsky, Inc.

AVIATION SAFETY CHIEF. Tuscaloosa, AL plant (2004-present). Officially cited for my initiative, dedication, and self-motivation, supervise teams of employees involved in the production of Apache helicopters.

- Was credited with saving time and money by establishing procedures for cleaning vital bolts, washers, and other components which previously had been replaced.
- Supervised and replaced GE 701-C engines, transmissions, gear boxes, and blades.
- Earned respect for sound judgment and contagious enthusiasm for meeting short notices and ensuring aircraft are maintained and ready for service at all times.
- Continue to emphasize safety in the work place with the result that my department has never suffered an accident or serious safety violation under my leadership.

TEAM LEADER & QUALITY ASSURANCE INSPECTOR. New Canaan, CT (2003-04). Trained and supervised eight people in an Apache helicopter construction facility with responsibility for timely completion as well as scheduled and unscheduled inspections.

- Controlled $250,000 in equipment in support of aircraft worth $16 million each.
- Provided 250- and 500-hour phase maintenance which included inspection as well as installation of major components. Performed strict quality assurance inspections according to precise specifications; became known as an astute problem solver.

Military experience: HELICOPTER MECHANIC. U.S. Army, Fort Rucker, AL, and Iraq (1999-03). Refined my technical, mechanical, and leadership skills while becoming known as a reliable professional who could be counted on, no matter how difficult the situation.

- Became proficient in such activities as removing, re-installing, and replacing helicopter engines, transmissions, blades, auxiliary power units, main rotor heads, Environmental Control Units, hydraulic components, and structure panels.
- Used tools, power tools, wrenches, special tools, and pneumatic tools while working with hoists and up to 10-ton cranes; became familiar with corrosion control.
- As custodian of the unit tool room, played a significant role in recognition with Army Aviation Association "Unit of the Year" and "Masters of Readiness" awards in 1999.

PERSONAL

Earned honors including U.S. Army Commendation and National Defense Service Medals and two Army Achievement Medals. Excellent work ethic and initiative.

Construction quality management

Date

Exact Name of Person
Title or Position
Exact Name of Company
Address
City, State, Zip

CONSTRUCTION QUALITY MANAGER
with experience as a project manager

Dear Exact Name of Person: (or Dear Sir or Madam if answering a blind ad.)

Thank you for your recent expression of interest in my experience. With the enclosed resume, I would like to make you aware of my interest in confidentially exploring management opportunities with your organization.

As you will see from my resume, since 2000 I have worked for Rocky Mountain Constructors, Inc., where I have been crosstrained as a Project Manager, Quality Control Manager, and Foreman. I have managed numerous projects and am currently the Project Manager for all U.S. government and City of Colorado Springs projects. I am experienced in managing projects including airport expansions as well as projects to establish and renovate utilities and roads. I am extremely experienced in managing construction projects on military bases, and I am accustomed to working with quality control technicians form the Corps of Engineers and other regulatory agencies such as OSHA. Proficient with all aspects of construction administration, I am skilled at negotiating contracts and subcontracts, creating master project schedules, preparing bids for government jobs, and writing purchase orders for materials. In my earliest job with Rocky Mountain Constructors, I was extensively trained as an Estimator and Quality Control Foreman.

The recipient of numerous letters of appreciation and awards for outstanding performance, I have earned respect for my emphasis on safety, my ability to motivate employees, and my strong bottom-line orientation. I have contributed significantly to my employer's bottom line. For example, I have improved net profit on government jobs from a 15% average to a 35% average. On one $1.6 million job which I managed, the company made a 48% profit.

I hope you will call or write me soon to suggest a time convenient for us to meet and discuss your current and future needs and how I might serve them. Thank you in advance for your time.

Yours sincerely,

Bradley Cameron

Alternate last paragraph:
I hope you will welcome my call soon to arrange a brief meeting to discuss your current and future needs and how I might serve them. Thank you in advance for your time.

BRADLEY CAMERON

1110½ Hay Street, Fayetteville, NC 28305 • preppub@aol.com • (910) 483-6611

OBJECTIVE
To benefit an organization that can use a construction industry Project Manager with previous experience in positions including Quality Control Manager, Foreman, and Estimator.

TECHNICAL KNOWLEDGE
Knowledgeable of OSHA and EMSHA.
Very knowledgeable of construction activities at Fort Carson and Peterson Air Force Base.

EDUCATION
Bachelor of Science (B.S.) in Business Administration, University of Colorado, Colorado Springs, CO, 2004.
Associate in Applied Science (A.A.S.), Pikes Peak Community College, Colorado Springs, CO, 2002.

TRAINING
Completed extensive training by military and civilian organizations including the following:
Asphalt Driving Workshop, Department of the Army
Construction Quality Management, U.S. Army Corps of Engineers
Project Productivity Improvements, Colorado AGC
Hot Mix Asphalt Construction, Department of Transportation

EXPERIENCE
CONSTRUCTION QUALITY MANAGER. Rocky Mountain Constructors, Inc., Colorado Springs, CO (2000-present). Began as an **Estimator**, and then was crosstrained in other aspects of the construction business. Worked as a **Laborer** and became familiar with all construction trades. Also worked as a **Quality Control Manager** and **Foreman.**
Highlights of projects: Have managed dozens of projects, and am the Project Manager for all U.S. government and City of Colorado Springs jobs.

- **Renovating and building roads:** For the City of Colorado Springs as well as for Ft. Carson, managed projects which involved the milling of asphalt on road, the raising and lowering of structures, establishing curbs, gutters, and sidewalks, and resurfacing roads.
- **Airport expansion:** Managed a $1.5 million expansion of the Colorado Springs Municipal Airport. Supervised 25 individuals who included foremen, pipe layers, asphalt grade crews, as well as striping and sealcoat crews. Built a general aviation ramp. Finished the project three months ahead of schedule and under budget.
- **Utilities:** Managed a $2 million project at Ft. Carson. Developed a site for the U.S. Special Operations, and was involved in establishing utilities such as water and sewer. Installed all utilities including water, sewer, storm drains, and erosion control. Supervised paving of surfaces. Managed a project which ran smoothly while supervising 35 people.
- **Projects at Fort Carson and Peterson Air Force Base:** Worked on numerous projects on military bases, and have become skilled at working with the government quality control standards. During one project at Peterson AFB, worked under extremely tight deadlines as the runway could be shut down for only three weeks. Built two taxiways and repaired an emergency sinkhole while also restriping the runway.
- **Extensive supervisory experience:** On numerous occasions, supervise 75 people including 10 foremen when the company's chief project manager is away.
- **Proficiency with government paperwork:** Have become proficient in writing subcontracts, writing purchase orders for materials, creating master project schedules using CPM, preparing bids for government jobs, and coordinating with quality control technicians from the Corps of Engineers.
- **Bottom-line results:** Have improved net profit on government jobs from a 15% average to a 35% average. The company made a 48% profit on one $1.6 million job.

PERSONAL
Excellent references on request. Have received numerous letters of appreciation and awards.

Well drilling operations

Date

Exact Name of Person
Title or Position
Name of Company
Address
City, State, Zip

CONSTRUCTION SAFETY SUPERVISOR

Dear Exact Name of Person (or Dear Sir or Madam if answering a blind ad):

With the enclosed resume, I would like to make you aware of my skills related to well drilling and heavy equipment operation while also introducing you to my reputation as a highly safety- and quality-conscious professional with outstanding supervisory skills.

As you will see from my resume, I recently added to my extensive credentials by earning the CDL Class "A" licenses with certifications including HAZMAT. While working for International Drilling Specialists, I have travelled throughout the world in order to work on heavy engineering and well drilling projects. While working on projects in Italy, Bolivia, Germany, and the U.S., I gained experience in all phases of equipment maintenance, operations, and supervision. I am licensed to operate dump trucks, backhoes, forklifts, scrapers, scoop loaders, and well drilling equipment as well as cargo trucks.

In my current position as Construction Safety Supervisor, I am known as a safety subject matter expert related to well drilling. I have overseen the timely completion of numerous projects and am respected for my decision-making skills and common-sense approach to problem-solving. In previous positions, I excelled in a variety of jobs in equipment operations and supervision while training both American and foreign personnel in heavy equipment operations. Well drilling teams that I have trained and managed have produced the best safety and quality assurance records ever achieved in the company's history.

If you can use a dedicated and mature professional with a versatile background and reputation for dependability and high personal standards, I hope you will call me soon to arrange a time when we might meet and discuss my qualifications and how I might contribute to your organization. I can provide excellent personal and professional references at the appropriate time.

Sincerely,

Francis Kessler

FRANCIS KESSLER

1110½ Hay Street, Fayetteville, NC 28305 • preppub@aol.com • (910) 483-6611

OBJECTIVE

To offer a strong background in construction and heavy equipment operations and management to an organization that can use a mature self-starter with experience and training in trucking as well as excellent skills as a supervisor, team leader, and trainer.

EDUCATION & TRAINING

Completing **National Training, Inc.,** three-week course for professional drivers seeking CDL Class "A" license with HAZMAT and other certifications.
A.A.S., Law Enforcement, Chaminade University of Honolulu, HI.
Received advanced instruction in leadership development courses, construction equipment operations, well drilling procedures, and combat lifesaving.

EQUIPMENT

Am licensed to operate heavy equipment including:

M-916 truck trailer	up to 8-ton backhoes	scoop loaders
up to 25-ton dump trucks	up to 10,000-lb. forklifts	scrapers and dozers
up to five-ton cargo trucks	Caterpillar 130G graders	LP-12 well drilling rigs

EXPERIENCE

Earned a reputation as a dedicated leader and supervisor, International Drilling Specialists:
CONSTRUCTION SAFETY SUPERVISOR. Honolulu, HI (2005-present). Became known as the "go to" person for technical safety guidance on well drilling procedures as supervisor of three people and controller of in excess of $1 million worth of equipment.

- Oversee the operation and maintenance of five well drilling equipment vehicles and associated materials.
- Cited for my emphasis on safety, ensured sites were prepared and work carried out in the most efficient, productive manner possible while maintaining zero accident rates.
- Was respected for my sound judgment and common sense decision-making displayed while helping subordinates overcome unforeseeable problems.
- Completed numerous special projects up to extremely high standards to include special projects in Ke'ehi Lagoon Park recreational area, a reconstruction project in Iraq, and an airfield construction project in Afghanistan.
- Expertly carried out additional duties which included ammunition handling, Safety Officer, and Master Driver for the section as well as Equal Opportunity Representative.

HEAVY EQUIPMENT SECTION SUPERVISOR. Bolivia (2003-04). During a nine-month project in Bolivia where we drilled wells under a contract with the country's leading public utility, provided leadership for an 11-person section while supervising the maintenance and operation of 13 pieces of heavy equipment valued in excess of $650,000.

- Trained Bolivian employees on equipment operations and well-drilling procedures.

WELL DRILLER. Germany (2001-02). During a special project in Germany, was praised for effectiveness in training seven German soldiers in all phases of well drilling and for my ability to produce outstanding results operating and maintaining a $250,000 system.

GENERAL CONSTRUCTION EQUIPMENT OPERATOR & WELL DRILLING SUPERVISOR. Italy (1998-01). Held multiple and varied duties which included operating and maintaining equipment such as nine-wheel rollers, vibratory rollers high-speed compactors, and trucks valued in excess of $400,000; supervising up to 20 people; and overseeing training and special projects.

- Was cited for my dedication to quality, concern for others, and skills as a trainer.

PERSONAL

Earned numerous honors including special medals awarded by foreign governments.

Automotive aftermarket

Date

Exact Name of Person
Title or Position
Exact Name of Company
Address
City, State, Zip

CONTINUOUS IMPROVEMENT MANAGER

Dear Exact Name of Person: (or Dear Sir or Madam if answering a blind ad.)

With the enclosed resume, I would like to make you aware of my interest in exploring employment opportunities with your organization. Although I am excelling in my current position and can provide excellent references at the appropriate time, I have made a decision that I wish to relocate back to Washington, DC, where I grew up.

As you will see from my resume, I have excelled for the past 12 years in a field which typically requires an engineering degree even though my B.A. from the Yale University in New Haven, CT, is in English Literature! Of course I have completed numerous technical management programs and extensive technical training.

After serving my country in the U.S. Navy and working in the Merchant Marines, I began employment with Eastern Industries as a Technical Writer, and I produced multiple users' manuals for people in disciplines ranging from accounts payable to purchasing and engineering. Because of my reputation as an effective communicator, I was then handpicked by Eastern Industries to work as a Project Manager. ISO 9000 was a new concept at that time, and my corporation needed a resourceful individual who could figure out how to make that concept work at multiple plants across the U.S. so that those plants could achieve the certification required by law. It was my job to visit those plants and provide the leadership for solving problems ranging from out-of-control scrap rates to manufacturing process problems. In my experience, I have learned that employees are the key to solving nearly all productivity and profitability problems.

In 2003 I was recruited by a plant within the Haven "family" to relocate to Maine to become its first Continuous Improvement Manager. Since arriving in Maine in 2003, I have played a key role in increasing plant profitability from 3% to 12%, and I have led efforts which have resulted in reducing inventory from $64 million to $28-32 million while improving product availability from 90% to 99%.

Although the work of improving internal processes is never complete, I have helped the plant to reach new record levels of profitability and efficiency, and I am proud that all plant employees have been trained in Continuous Improvement. A native of DC, I now am yearning to "go back home" and I am selectively exploring the possibility of joining an organization which can use a creative problem solver, strong leader, and skilled efficiency expert. If you feel that you have internal problems which you would like to discuss with me or are not even sure if you have a full-time job for a problem solver like myself, I still would enjoy meeting you and confidentially discussing your needs.

Yours sincerely,

Dylan Cooper

DYLAN COOPER

1110½ Hay Street, Fayetteville, NC 28305 • preppub@aol.com • (910) 483-6611

OBJECTIVE

I want to contribute to the profitability and efficiency of an organization that can use a skilled problem solver who offers technical expertise related to the areas of quality assurance systems, manufacturing systems, and general continuous improvement management.

EXPERIENCE

Advanced in the following track record of promotion with Haven Industries, a publicly traded conglomerate headquartered in New Haven, CT; Eastern Corporation purchased assets from Haven Industries in 2004 and I have continued with Eastern Corporation in an expanded role since the purchase:

2003-present: CONTINUOUS IMPROVEMENT MANAGER. EastHouse Division of Eastern Corporation. Was aggressively recruited by a plant in the Haven "family" after I acted as an internal consultant.

- Played a key role in increasing the plant's profitability from 3% to 12%, nearly unheard of in the automotive aftermarket industry. Led efforts which resulted in reducing inventory from $64 million to $28-32 million while boosting product availability from 90% to 99%. Have provided strong leadership through a vibrant period of change.
- Manage four people directly; established employee leadership teams.
- Created a dynamic employee team approach within the plant which has resulted in significant contributions to strategic planning, technical problem solving, and employee morale. Established the site's first Continuous Improvement (CI) Team as well as the Continuous Improvement Committee which is comprised of employees across multiple disciplines; developed the 20 Keys of Continuous Improvement Program; set up the AMPS Performance Excellence and Business Excellence Systems.

1997-03: PROJECT MANAGER. New Haven, CT. Became the company's foremost authority on ISO 9000/QS 9000 and quality assurance systems; succeeded in obtaining QS 9000 certification for all automotive OE plants within one calendar year.

- In an essentially entrepreneurial role, was handpicked by the company to implement new QS 9000 guidelines for the Automotive Group. Implemented the company's first ISO 9000 at a Maine plant, and established the model for future implementation.

1996-97: PROJECT MANAGER. New Haven, CT. Was specially selected for this newly created position by the corporation's new Director of Total Quality Management.

1995-96: TECHNICAL WRITER. New Haven, CT. Interviewed experts in Material Requisition & Planning (MRP), Purchasing, Engineering, Accounts Payable, and Engineering.

Other experience:

NAVAL SEAMAN. Merchant Marines, locations worldwide. Developed strong technical problem-solving skills while operating a large seagoing vessel and performing complex maintenance aboard ship; managed and trained mechanics and maintenance professionals.

NAVIGATION QUARTERMASTER. U.S. Navy, locations worldwide. Was promoted ahead of my peers while serving in the U.S. Navy; received outstanding evaluations.

EDUCATION

Completed **B.A. in English Literature,** Yale University, New Haven, CT, 1991.
Extensive executive development related to Quality Assurance Systems, Manufacturing Systems, Continuous Process Improvement, ISO 9000, ISO 14000, and QS 9000.

PERSONAL

Highly intuitive and analytical problem solver who believes that employee involvement is essential to solving most profitability, process, efficiency, and other problems.

Telecommunications engineering

Date

Exact Name of Person
Title or Position
Name of Company
Address
City, State, Zip

COORDINATOR OF QUALITY ASSURANCE (QA) OPERATIONS

Dear Exact Name of Person: (or Dear Sir or Madam if answering a blind ad.)

I would appreciate an opportunity to talk with you soon about how I could contribute to your organization through my experience and skills as well as through my reputation as an adaptable quick learner with a talent for relating to and working with others both as a team member and in supervisory roles.

As you will see from my resume, I offer a "track record" of successful performance. My time with Black & Decker has been in the field of Quality Assurance for the Mobile Subscriber Equipment (MSE) system installations and retrofit projects. Earlier I served my country in the U.S. Army while gaining experience in the areas of inventory control and logistics as well as supervising the operation of a diverse inventory of telecommunications equipment.

I feel that I offer a well-rounded background which includes planning, scheduling, supervising, training, and evaluating as well as records administration. I feel that this versatility would allow me to easily and quickly step into a variety of positions for a company which could benefit from my many skills.

I hope you will welcome my call soon to arrange a brief meeting at your convenience to discuss your current and future needs and how I might serve them. Thank you in advance for your time.

Sincerely yours,

Anita Goodey

Alternate last paragraph:
I hope you will call or write me soon to suggest a time convenient for us to meet and discuss your current and future needs and how I might serve them. Thank you in advance for your time.

ANITA GOODEY

1110½ Hay Street, Fayetteville, NC 28305 • preppub@aol.com • (910) 483-6611

OBJECTIVE

To apply my expertise in quality assurance and supply operations to an organization seeking an adaptable, detail-oriented professional with outstanding planning skills.

EXPERIENCE

Earned advancement while gaining a well-rounded background of experience in this "track record" of achievements with Black & Decker Corp.:

COORDINATOR OF QUALITY ASSURANCE (QA) OPERATIONS. B&D Regional Support Center, Lawrence, KS (2005-present). Achieved a "perfect error-free" evaluation during my first audit in this location in a job which included a wide range of functions such as inspecting and resolving computer software problems.

- Reduced equipment calibration costs by 30-40% in this facility which is the repair/maintenance shop for Mobile Subscriber Equipment supplied to the U.S. Army.
- Conducted audits to confirm contract compliance and made daily checks of calibrated and warranty items, ensuring that test equipment was available and accurate.
- Applied my communication skills teaching CPR and safety classes to up to 30 people.
- Supervised each phase of quality assurance documentation and recordkeeping.
- Inspected incoming and outbound shipments to ensure proper packaging.

QUALITY ASSURANCE INSPECTOR. Various locations (2003-04). Received a "Superior Performance Award" for my efforts while traveling throughout the U.S. and to Europe on a retrofit team which was upgrading equipment sold to the U.S. Army.

- Earned praise for my willingness to work the long hours required for project success.
- Conducted inspections when equipment was received for repair or maintenance, monitored all stages of work, and prepared the QA documentation. Constantly reviewed and updated QA plans and procedures; reviewed the final configuration packages.

QUALITY ASSURANCE (QA) INSPECTOR. Fort Stewart, GA, Fort Drum, NY, and San Francisco, CA (2001-03). Was promoted to the retrofit team on the basis of my performance and demonstrated ability to establish and maintain rapport with customers. Maintained records on system failures; prepare corrective actions and report on them.

Gained experience in the areas of supply and telecommunications operations, U.S. Army:

SUPERVISORY SUPPLY SPECIALIST. Germany (1999-00). Ensured that the members of a 200-person organization had adequate supplies backed by complete and accurate records.

SUPERVISORY TELECOMMUNICATIONS OPERATOR. Fort Hood, TX (1997-99). Scheduled, conducted, and recorded preventive maintenance actions.

- Reduced the inventory of non-operational equipment by an impressive 90% through the establishment of production control and quality assurance procedures.

EDUCATION & TRAINING

A.A. degree, Central Texas College, Killeen, TX, 1995.

Completed military and corporate training programs in radio communications operations, effective customer service skills, and leadership and supervisory effectiveness.

TECHNICAL KNOWLEDGE

Learned to operate a variety of communications equipment including multichannel radios; switching gear; and VHF, UHF, AM, and FM receivers and transmitters.

Have a working knowledge of Microsoft Word, Excel, PowerPoint, and Adobe PageMaker.

PERSONAL

Held a Secret security clearance. Offer a reputation for being an enthusiastic and outgoing individual. Am available for relocation and travel.

Courtroom safety and security

Date

Exact Name of Person
Title or Position
Name of Company
Address
City, State, Zip

COURT SAFETY OFFICER

Dear Exact Name of Person: (or Dear Sir or Madam if answering a blind ad.)

With the enclosed resume I am formally indicating my interest in the job of United States Marshal for the Western District of Arizona, and I am requesting that you support me for this position. I can tell you that you would be backing a respected professional with an unblemished record of service to my country along with a thorough knowledge of what is required of the U.S. Marshal.

As you will see from my resume, I have worked for the United States Marshals Service (USMS) as a Court Security Officer and, during that time, I have established excellent working relationships with individuals from the USMS, other law enforcement agencies, U.S. attorneys, and the judiciary. Previously I worked for the Arizona Wildlife Commission, advancing into a supervisory position which involved training and managing other wildlife officers while enforcing state and federal laws related to hunting, fishing, and boating.

I strongly believe that the U.S. Marshal can make a difference in the Western District, and I believe he will make a difference not by sitting behind a desk but by a "hands-on" approach to working with, helping, and motivating deputies. An energetic "hands-on" professional with a "walk around" management style, I know all the deputies, judicial officials, and most of the law enforcement officers in the district and could count on their 100% support.

Please support me for this position, and let me hear from you if you need any other supporting documentation from me. Thank you very much.

Sincerely yours,

Arthur C. Bisque

ARTHUR C. BISQUE

1110½ Hay Street, Fayetteville, NC 28305 • preppub@aol.com • (910) 483-6611

OBJECTIVE

To serve as the United States Marshal for the Western District of Arizona.

EXPERIENCE

COURT SAFETY OFFICER. United States Marshals Service (USMS), Tempe, AZ (2000-present). Have developed excellent working relationships with individuals from the Marshals Service, law enforcement agencies, U.S. attorneys, and the judiciary while involved in a wide range of activities related to providing security and protection.

- ***entrance control:*** Operate and enforce a system of personal identification which includes checking handbags, packages, and other items to detect weapons and contraband.
- ***roving patrol:*** Conduct roving patrols of the court area in accordance with schedules.
- ***fixed post:*** Maintain a fixed, stationary position outside and inside the chambers of courtroom judges and jury rooms in order to prevent unauthorized entrance.
- ***personal escort:*** Provide a personal escort for judges, court personnel, attorneys, jurors, and witnesses when directed to do so in order to assure their personal safety.
- ***law and order:*** Am responsible for the detection and detention of any person(s) seeking to gain unauthorized access to court proceedings.

AREA SERGEANT. Arizona Wildlife Commission, Tempe, AZ (1990-00). Refined my leadership ability and problem-solving skills while supervising and directing several wildlife officers in a geographical area in the enforcement of AZ and federal laws related to hunting, fishing, other game activities, and boating.

- ***employee training/supervision:*** Trained numerous officers in the wildlife field; evaluated their performance.
- ***organizing/scheduling/coordinating:*** Set up work details to handle wildlife activities; worked with state and federal law enforcement agencies; assisted in search and rescue missions.
- ***inventory control:*** Procured new equipment and monitored its maintenance and care.
- ***budgeting and finance:*** Planned and administered budgets of varying sizes.

WILDLIFE ENFORCEMENT OFFICER. Arizona Wildlife Commission, Tempe, AZ (1985-90). Enforced game, fish, and boating laws in an assigned area and assisted other officers in high violation locations; maintained equipment in top condition.

EDUCATION & TRAINING

Court Security Officer School (law enforcement training), 2000.
Institute of Government (in-service training), Tempe, AZ, 1998.
Arizona Justice Academy (AZ Criminal Code), Salemburg, AZ, 1995.
Riot and Crowd Control, Tempe, AZ, 1990.
Institute of Government (leadership training), Tempe, AZ, 1989.
Coast Guard Boarding School, San Diego, CA, 1988.
Basic, Intermediate, and Advanced Law Enforcement Courses, 1985-present.

DISTINCTIONS & HONORS

- Received letters of commendation from a federal judge for outstanding work
- Received letter of commendation for exceptional performance from U.S. Attorneys Office
- Was the recipient of numerous awards from clubs and civic organizations
- Received the State Conservation Award
- Was named Officer of the Month

PERSONAL

Have a strong desire to strengthen law enforcement in the Western District of Arizona. Can pass the most rigorous security background check; background is free of obscurities.

New product development

Date

Exact Name of Person
Title or Position
Name of Company
Address
City, State, Zip

DIRECTOR, TEST DIRECTORATE

Dear Exact Name of Person: (or Dear Sir or Madam if answering a blind ad.)

Can you use an experienced executive who has demonstrated an ability to improve organizational effectiveness while implementing proven principles of Total Quality Management (TQM) at all levels and in all functional areas?

Extensive management experience
As you will see from my resume, I was promoted to the rank of colonel in the U.S. Army in 2001 and have thoroughly enjoyed serving my country while excelling in a variety of line management and top-level strategic planning/consulting roles. Very early in my career I was selected as a first lieutenant for a "company commander" job normally held by a captain, and I "turned around" an organization which had failed its annual inspection. Later on I was selected for one of the military's most prestigious management positions as "battalion commander" (chief executive officer) of a 750-person airborne infantry organization. In every job I have held, I have been commended for my ability to instill a positive, optimistic, and enthusiastic attitude in employees. In 2001 I was handpicked as Chief of Staff of an 1,800-person Joint Task Force providing humanitarian relief to 14,000 migrants, and I received a special award from the Chairman of the Joint Chiefs of Staff for my accomplishments.

Top-level experience related to safety and quality assurance
My analytical and problem-solving skills have also been tested in jobs requiring expertise related to safety and quality assurance. In my current job as Director of a U.S. Army Test Directorate, I am the senior manager of a testing organization with 60 civilian and 40 military employees conducting tests on products, systems, procedures, and equipment Army-wide. In my prior position, I served for three years as Inspector General and was the chief "watchdog" directing investigations of fraud, waste, abuse, and mismanagement at four military installations. I was commended for managing a safety and quality assurance programs cited as "the best Army-wide." In a previous job I applied my safety and quality assurance knowledge in a job which placed me in charge of supervising a sophisticated 15-person staff conducting analyses of the effectiveness of training centers in the U.S., Europe, and other countries.

I can provide outstanding personal and professional references at your request, and I hope you will call or write to suggest a time when we might meet to discuss your current and future needs and how I might serve them. Thank you in advance for your time.

Sincerely yours,

James Koontz

JAMES KOONTZ

1110½ Hay Street, Fayetteville, NC 28305 • preppub@aol.com • (910) 483-6611

OBJECTIVE

To offer the executive abilities I have gained in a distinguished career as a military officer to an organization that can use a creative leader with expert knowledge of Total Quality Management which I have applied while managing people, finances, programs, and projects.

EDUCATION

M.A. degree, Management, Central Texas University, Fort Bragg, NC, 1991.
B.A. degree, History, American University, Washington, D.C., 1980.
Quality Control: Completed formal and hands-on training in Total Quality Management (TQM) and Total Quality Advantage (TQA).

EXPERIENCE

DIRECTOR, TEST DIRECTORATE. U.S. Army, Fort Benning, GA (2005-present). Was specially selected for this position as senior manager of a testing organization with 60 civilian and 40 military employees involved in planning, implementing, and reporting on the performance of tests performed Army-wide on:

new procedures
equipment systems
transport inside aircraft
transport external to aircraft
special operations
classified matters
materiel for personnel and equipment parachute drop

- On my own initiative and utilizing my background in Total Quality Management, conducted a complete inventory in my first 30 days on the job (the organization's first-ever inventory) and identified $135,000 of excess equipment which we sold.
- Maintained quality test production while reducing the workforce by 20%. Provided oversight for a $6M budget while managing $6.5M in property and $17M in equipment.
- Improved morale and productivity of the workforce by improving the flow of information, implementing a new training program, and initiating merit-based awards.
- Have played a key role in fielding many of the military's new products and systems, and am considered one of the Army's experts on airdrop capabilities of aircraft.
- Hold Top Secret security clearance based on Special Background Investigation (TS/SBI).

INSPECTOR GENERAL. U.S. Army, Fort Campbell, KY (2002-05). Because of my reputation for integrity and my proven problem-solving skills, was handpicked for this top-level "watchdog" position directing investigations of fraud, waste, abuse, and mismanagement at four military installations, including the world's largest U.S. military base.

- Saved hundreds of thousands of dollars through investigative activities while also implementing new Total Quality Management (TQM) procedures and articulating the philosophy that the highest standards of conduct, ethics, and integrity were acceptable.

CHIEF OF OPERATIONS. U.S. Army, Fort Campbell, KY (1999-01). Revitalized the Emergency Operations Center, improved training programs, created an atmosphere of teamwork, and was praised for "placing organizational interest ahead of personal convenience."

Highlights of other experience:
TRAINING CHIEF. Played a major role in shaping Army policy with regard to training, organization, equipment, and management in this job conducting analyses of the effectiveness of training centers.
COMPANY COMMANDER. Was selected as a first lieutenant for a job normally held by a captain; excelled as "chief executive officer" of a 150-person infantry organization composed of both German and U.S. soldiers; assumed command after the organization had failed to pass its annual inspection and then instituted changes which earned praise from inspectors.

Vehicle rental operations

Date

Exact Name of Person
Title or Position
Name of Company
Address
City, State, Zip

DISTRICT MANAGER FOR SAFETY AND MAINTENANCE ADMINISTRATION

Dear Exact Name of Person: (or Dear Sir or Madam if answering a blind ad.)

I would appreciate an opportunity to talk with you soon about how I could contribute to your organization through my background in customer service, personnel and facility management, and safety in the field of truck leasing and rental operations.

During my years with a national rental organization, I handled customer and driver relations, managed maintenance facilities, developed and provided safety and maintenance instructional programs, and was responsible for assisting sales managers and customers. I was also in charge of environmental issues as well as Department of Transportation regulations. I am knowledgeable of all regulations promulgated by the Federal Highway Administration, OSHA, and other regulatory agencies.

As you will see from my resume, I am skilled in communicating with personnel ranging from customers and drivers to maintenance mechanics and service managers, and I have been heavily involved in conducting training programs and drivers' meetings. I am always looking for ways to improve safety, efficiency, and customer service. One example of my strong leadership in the safety area is my involvement in "turning around" the Florida district from last place in the nation's 118 districts to "first in the nation" in percent-of-profit-to-revenue in a five-year period. I have also led the organization to achieve its best safety results during independent audits of company operations. For example, I have established safety programs which reduced employee accident rates to an all-time low, and I have worked extensively with regulators to satisfy their concerns about the disposition of the company's underground storage tanks.

I hope you will welcome my call soon to arrange a brief meeting at your convenience to discuss your current and future needs and how I might serve them. Thank you in advance for your time.

Sincerely yours,

Timothy Canters

Alternate last paragraph:
I hope you will call or write soon to suggest a time convenient for us to meet and discuss your current and future needs and how I might serve them. Thank you in advance for your time.

TIMOTHY CANTERS

1110½ Hay Street, Fayetteville, NC 28305 • preppub@aol.com • (910) 483-6611

OBJECTIVE

To benefit an organization that can use a well-rounded professional who offers specialist knowledge of heavy truck maintenance management along with "hands-on" expertise in inventory control, safety, purchasing, sales, and customer service.

EXPERIENCE

Advanced in this "track record" with the U-Haul Company rental chain at various locations in Florida, Georgia, and Alabama:

DISTRICT MANAGER FOR SAFETY AND MAINTENANCE ADMINISTRATION. Panama City, FL (2005-present). Control the maintenance programs of four branch managers in six facilities with a total of 60 employees and 600 vehicles while overseeing safety matters.

- Established safety programs which reduced employee accident rates while also handling environmental concerns related to underground storage tanks.
- Coordinate arrangements to centralize purchasing activities for parts, tires, tools, and equipment as well as safety training. Provide customers with advice on the Federal Highway Administration's latest rules and regulations and other information.

DISTRICT MAINTENANCE MANAGER. Panama City, FL (1994-05). "Turned around" an unprofitable district which covered the northwestern part of Florida and included more than 600 class 6,7, and 8 trucks and trailers and 60 employees in as many as eight maintenance facilities.

- Brought the district from last place out of 118 districts nationwide to earn the **number one position** during the five years from 1994 to 1999; and remained in the nation's top 20 from 1999 on. Achieved $16 million in revenue through vastly improved procedures including substantial maintenance expense reductions: cut costs from above 28% of revenue to below 20%.
- Directed numerous operational areas including purchasing, training, sales support, and customer and driver relations. Established a safety program which led to an impressive reduction in accidents from 25 in 2001 to only nine in 2002.

Highlights of other experience:

- Provided technical advice and troubleshooting skills for the Georgia-Florida region which contained 3,000 pieces of equipment, 30 maintenance facilities, 400 employees, and earned approximately $40 million annually, **REGIONAL MAINTENANCE MANAGER**, Savannah, GA.
- As **DISTRICT SERVICE MANAGER**, introduced changes which reduced employee turnover and maintenance costs in a $6 million annual revenue three-facility district.
- Was credited with improving the company's image as the **BRANCH SERVICE MANAGER** in Gainesville, FL.
- Became the company's youngest **BRANCH SERVICE MANAGER** at age 24 and physically turned an abandoned building into the first U-Haul facility in Pensacola, FL.

EDUCATION

Completing Bachelor of Science in Industrial Safety, Panama City University, Panama City, FL; am pursuing degree in my spare time and expect to graduate in 2006.

Completed over 200 hours of U-Haul corporate-sponsored training programs in the areas of customer and driver relations, personnel management, preventive maintenance, and diesel engine electronics controls.

Studied Auto/Diesel Welding at the Gulf Coast Community College, Panama City, FL.

PERSONAL

Extensive background in customer service, personnel relations, and facility management. Am a proven motivator and communicator with a talent for team building.

Waste water treatment plants

Date

Exact Name of Person
Title or Position
Name of Company
Address
City, State, Zip

ELECTRICAL SUPERINTENDENT

Dear Exact Name of Person: (or Dear Sir or Madam if answering a blind ad.)

With the enclosed resume, I would like to make you aware of my interest in exploring employment opportunities with your organization. I am an Electrical Superintendent with extensive experience in safety and quality assurance related to waste water treatment plants.

Expertise as an electrical superintendent
As you will see from my resume, I have excelled as an electrical superintendent with Atlantic Industrial Constructors since 2002. While specializing in pumping station upgrades, I have acted as electrical superintendent for projects with the city of Richmond as well as counties including Richland, Cumberland, and others. I am accustomed to assuming complete profit-and-loss responsibility on projects, and typically I stage the job, order materials, hire workers, and handle my own scheduling. The company has never lost money on any project I have managed, and I am skilled at bringing a job in on time and within budget. In prior experience as a self-employed Electrical and HVAC Contractor in North Carolina, I worked on both residential and commercial jobs, and I had a contract with the State of North Carolina to work on span bridges. I held an Unlimited Master Electrical License in the State of North Carolina by examination. I have an outstanding safety record, and I am experienced in working with onsite inspectors.

Highlights of previous experience
While serving my country in the U.S. Army, I held a Top Secret security clearance and supervised maintenance of a missile system in Europe. Subsequently I worked as a Commercial Refrigeration Technician prior to becoming an Electrical and HVAC Contractor. Even as a youth, I was trained in the electrical business since my dad was an electrical contractor for federal work. I was trained to perform simple electrical tasks as a young boy, and that sparked my interest in the field.

If you can use a skilled problem solver and troubleshooter with extensive project management expertise, I hope you will contact me to suggest a time when we might meet to discuss your needs. I can provide outstanding references at the appropriate time.

Yours sincerely,

Matthew Fisher

MATTHEW FISHER
1110½ Hay Street, Fayetteville, NC 28305 • preppub@aol.com • (910) 483-6611

OBJECTIVE

To benefit an organization that can use an accomplished Safety Superintendent with an excellent safety record along with a background in project management and quality control.

EXPERIENCE

ELECTRICAL SAFETY SUPERINTENDENT. Atlantic Industrial Constructors, Inc., Richmond, VA (2002-present). Began employment with Cavuto Rigging Co. and continued with Atlantic Industrial Constructors when they bought out Cavuto in 2002. Have been commended for outstanding safety procedures. I assume total responsibility for the electrical projects I manage; the company hands me the plans and then I stage the job, order materials, hire workers, and perform the job while handling my own scheduling. Am skilled in working with onsite inspectors and engineering inspectors for project completion. Projects included:

- **City of Columbia Pump Station upgrade.** Upgraded pumping stations and installed generators at 13 different locations.
- **County W.T.P upgrade.** Upgraded a plant from five to 10 million gallons a day.
- **Virginia Beach Emergency Shelters.** After Hurricane Kaylee devastated areas in VA, undertook a complex project to turn four high schools into Virginia Beach Emergency Shelters. Supervised five electricians; completed the project while school was in session.
- **Relocation of centrifugal welders from American Electric.** AE put me in charge of disassembling a centrifugal welder which made jet engine rotors for 747s from Lexington, KY to Raleigh, NC to disassemble it and restore it to operating condition.
- **Relocation and reassembly of plant equipment.** Disconnected the equipment of manufacturing plants in Miami to New York which made metal projects for the construction industry for the T.P Corporation. Transported and reassembled equipment.
- **Richland County, SC pumping stations.** Two contracts for pumping stations.
- **Myrtle Beach water treatment plant upgrade.** Managed five electricians while upgrading the Myrtle Beach plant from four million to eight million gallons a day.
- **Main Road pumping station.** Upgraded the capability of this pumping station to store purified water; installed electrical instrumentation.
- **Charleston County water treatment plant upgrade.** Modified and modernized equipment at a water treatment plant.
- **Wilmington waste water treatment plant upgrade.** Boosted capacity.

Other experience:
COMMERCIAL REFRIGERATION TECHNICIAN. Fred Meyer, AK. Set up a preventive maintenance program for a small chain with five stores in different geographical locations.
SELF-EMPLOYED ELECTRICAL AND HVAC CONTRACTOR. Became proficient with gas packs and oil furnaces. Worked on residential and commercial jobs. Wired houses, schools, and performed a contract with the State of North Carolina for electrical work on span bridges. Worked on fire alarms, intercoms, and security systems.

EDUCATION & LICENSES

Certified Technician I and II, Downeast Training & Certification, since 2002.
Completed the nine-month U.S. Army school for electricity, hydraulics, pneumatics, and diagnostics.
Technician I and II, Refrigeration, since 1990.
Held Unlimited Master Electrical License in the State of North Carolina by examination.

SKILLS

I have developed versatile skills in the electrical field since my youth. My dad was an electrical contractor for federal work, and he taught me the importance of strict adherence to specifications. Have wired houses, schools, and plants. Adept at managing multiple high-priority projects. Excellent references available. Outstanding safety record.

Government contracts and new products

Date

Exact Name of Person
Title or Position
Name of Company
Address
City, State, Zip

ELECTRONIC PRODUCT ASSURANCE MANAGER

Dear Exact Name of Person: (or Dear Sir or Madam if answering a blind ad.)

With the enclosed resume, I would like to make you aware of my experience in the area of communications/RADAR electronics with special expertise related to computer networking, quality assurance, test equipment, electronics product development, and contract management.

I was handpicked for my current position as Electronic Product Assurance Manager from among the Air Force's best and brightest technical experts. I have advanced to increasing responsibilities, and I currently perform technical evaluations of contractor proposals valued in excess of $150 million. I am also directing the Product Assurance Program for the RC-135 Rivet Joint aircraft. While overseeing the operation of four repair shops, three of which are overseas, I correct logistics problems and maintain separate equipment accounts for each shop with a total in excess of $200 million. I continuously advise project engineers and oversee contractor progress on modifications to the aircraft and ground simulators.

I have become widely respected for my initiative, resourcefulness, and exceptionally strong troubleshooting and problem-solving skills. On one occasion, I identified and then corrected 25 software anomalies which would have prevented system acceptance of a new ground simulator. I have saved the Air Force millions of dollars through designing new test procedures and resolving stubborn product deficiencies and software deficiencies. On one formal performance evaluation, I was described as "never missing an opportunity to improve a process, increase maintainability/reliability, or cut costs."

Having completed extensive training, I possess the technical knowledge and ability needed to efficiently perform troubleshooting, installation, design testing, and evaluation work. Frequently, I evaluate and lab test products and software packages to determine possible malfunctions and recommend appropriate solutions. In addition to using laboratory facilities to evaluate, test, and provide solutions to reported technical problems, I have conducted formal technical training on all supported products. In previous positions, I excelled in jobs as an Instructor and Quality Assurance Manager. I hold a Top Secret (TS/SBI) security clearance.

If you can use a top-notch technical expert with the ability to think strategically and solve tough technical problems, I hope you will contact me to suggest a time when we might talk about your needs. I can provide outstanding references at the appropriate time.

Sincerely,

Jonathan Curry

JONATHAN CURRY

1110½ Hay Street, Fayetteville, NC 28305 • preppub@aol.com • (910) 483-6611

OBJECTIVE

To offer an organization my information technology expertise related to computer networking, quality assurance, electronic systems development, and contract management.

TRAINING

Completed computer educational services and training programs related to Customized Sybase Training, System Administration, and Operations of the UNIX System.

Completed U.S. Air Force training programs related to electronics, computer operations, quality assurance, network design, and computer systems.

CLEARANCE & COMPUTER EXPERTISE

Top Secret security clearance (TS/SBI)

Experienced in directing and participating in the analysis, design, evaluation, testing, and quality assurance of computer hardware and software systems.

- General knowledge of all types of test equipment.
- Experienced in reading engineering and mechanical schematics, drawings, and wiring diagrams; performed hardware/software deficiency testing and evaluations.

EXPERIENCE

ELECTRONIC PRODUCT ASSURANCE MANAGER. U.S. Air Force, Pope AFB, NC (2005-present). Was specially chosen for this job among the Air Force's top technical experts; direct the Product Assurance Program for the RC-135 Rivet Joint aircraft (RJ), and advise project engineers overseeing contractor progress on modifications to the aircraft.

- Perform technical evaluations of contractor proposals valued in excess of $150 million; led six different contractor teams in the modification and testing of aircraft systems.
- Through my resourcefulness in utilizing commercial and open system architecture, was credited with saving the Air Force $200,000 in software development. Designed new test procedures which reduced repair cost and increased equipment availability 45%.
- Revitalized the organization's quality deficiency reporting and analysis system; cleared a backlog of 20 hardware quality and 15 software deficiency reports.
- Created a new maintenance manual for the RJ Electronic systems which increased system maintainability by expanding system troubleshooting procedures.
- Am considered the "point of contact" Air Force-wide on the state-of-the-art electronic systems used on strategic/tactical reconnaissance aircraft, special purpose aircraft, and ground processing systems.
- Identified and then corrected 25 software anomalies which would have prevented system acceptance of a ground simulator. Managed the delivery and configuration of new UNIX workstations; trained new personnel, and put the equipment to work ahead of schedule.
- Spearheaded efforts to upgrade the Condor Hawk system; prevented system incompatibilities by updating 258 tuners in preparation for the new resource controller; conducted final acceptance testing for a $1 million test station for the Condor system, and corrected software discrepancies that would have prevented full utilization.

INSTRUCTOR & QUALITY ASSURANCE MANAGER. U.S. Air Force, Andrews AFB, MD (2003-05). Evaluated in-flight maintenance technicians on two version aircraft.

- Wrote and revised open-book qualifications and closed-book emergency procedures tests.
- Coordinated changes to systems/computer checklists and created new checklists for new systems; instructed personnel in the operation, malfunction analysis, and repair of electronic warfare receivers, and operations of support test equipment.
- Credited with locating a major system malfunction on a new multimillion-dollar initial acceptance flight, enabled contractors to correct the firmware problem before delivery.

PERSONAL

Awarded Distinguished Flying Cross and other medals. Outstanding references on request.

Pollution of air and water

Date

Exact Name of Person
Title or Position
Name of Company
Address
City, State, Zip

ENVIRONMENTAL COMPLIANCE INSPECTOR

Dear Exact Name of Person: (or Dear Sir or Madam if answering a blind ad.)

With the enclosed resume, I would like to make you aware of my interest in exploring employment opportunities with your organization. I have recently relocated to South Carolina with my husband because of his new position, and I have much to offer an organization that can use a versatile hard worker with strong communication, management, computer operations, and budget skills. I am interested in the management position we recently discussed on the telephone.

As you will see from my resume, I have worked in the environmental compliance field and am the only Environmental Compliance Inspector for my employer in Dallas, TX. I have acted as spokesperson for environmental issues and was involved in training employees at all levels. Continuously involved in quality assurance activities, I conducted announced and unannounced inspections to determine compliance with regulations. I played a major role in convincing the company to upgrade its practices related to the disposal of toxic byproducts before they pollute the air or water.

Although I have excelled in my recent job, I have decided to explore opportunities outside the environmental field at this next phase in my career. You will notice from my resume that I offer strong computer knowledge. I am skilled in database management using Microsoft Access, and I am proficient with all Microsoft programs. I have worked effectively in situations where I was a bookkeeper, budget assistant, computer program analyst, and administrative assistant.

One of my main strengths is my ability to adapt easily to new environments as I have a natural problem-solving orientation. Known for my attention to detail in all matters, I excel in situations that require excellent analytical and problem-solving skills. I also offer the ability to act as a spokesperson for an organization utilizing the public speaking and problem-solving skills which I have refined through experience.

If you can use a versatile and adaptable professional with knowledge in numerous areas to become a part of your team, I hope you will contact me to suggest a time when we might meet in person to discuss your needs. I can provide outstanding references at the appropriate time.

Sincerely,

Amy Vanderbilt

AMY VANDERBILT

1110½ Hay Street, Fayetteville, NC 28305 • preppub@aol.com • (910) 483-6611

OBJECTIVE

To benefit an organization that can use an articulate, experienced professional with exceptional planning and organizational skills who offers a versatile background related to customer service, public relations, bookkeeping, and project management.

EDUCATION & CERTIFICATION

Completed two years towards a B.S. in Business Management, University of Virginia. Attended numerous courses related to Environmental Laws and Regulations, Hazardous Material Waste Handling, and CPR/First Aid.

COMPUTERS

Completed formal coursework for Microsoft applications and the Windows operating system. Knowledgeable of Windows and Microsoft Access, Excel, and Word.

- Skilled in database management using Microsoft Access.
- Have created presentations with PowerPoint.

EXPERIENCE

ENVIRONMENTAL COMPLIANCE INSPECTOR. Bristol Corporation, Dallas, TX (2000-present). Advanced from Environmental Assistant (2000-04) to Environmental Compliance Inspector (2005-present). Resigned in order to relocate with my husband.

- As the only inspector in Dallas for this major corporation, conducted announced and unannounced inspections of facilities to determine compliance with environmental regulations; provided guidance and made recommendations for corrective action, then prepared formal reports of the inspection.
- On my own initiative, updated the environmental compliance checklist/inspection form which was four years out of date. Played a key role in shaping the company's rigorous new policies related to disposing of toxic products before they pollute the air and water.
- Was extensively involved in training personnel at all levels, from executives to entry-level personnel; conducted classes and briefed executives and personnel.
- Acquired expertise in quality assurance and environmental compliance inspection.
- Have acquired vast expertise related to environmental issues while serving in a highly visible capacity as the spokesperson for environmental compliance.
- Received the highest possible evaluations on all performance appraisals; was praised for "consistently exercising sound judgment," for my "hard work in improving compliance."

COMPUTER ANALYST. Department of Defense, Ft. Hood, TX (2000). Handled data entry and computer systems support for a project that computerized inventory data for eight schools.

ADMINISTRATIVE ASSISTANT. Department of Defense, Ft. Dix, NJ (1998-99). Maintained a student database using Excel while also handling purchase requisitions and assisting in contract modifications; worked closely with the Public Affairs office, and was Acting Public Affairs Officer in the absence of the Director.

BUDGET ASSISTANT. Womack Army Community Hospital, Ft. Bragg, NC (1998). Managed expenditures of the Supplemental Care Program utilizing automated accounting system, and performed audits of medical facility treatment activities and pharmacy inventory.

BUDGET ASSISTANT. U.S. Army, Belgium (1995-97). Maintained ledgers of fund distribution for major units while gathering data for budget analysts, confirming accuracy of reports, as well as reconciling and balancing ledgers.

PERSONAL

Have earned three Sustained Superior Performance Awards. Known as a service-oriented professional with strong analytical and problem solving skills. Excellent references available.

Raw food inspection

Date

Exact Name of Person
Title or Position
Name of Company
Address
City, State, Zip

FOOD SAFETY INSPECTOR

Dear Exact Name of Person: (or Dear Sir or Madam if answering a blind ad.)

I am writing to express my strong interest in the position as Safety Manager which you recently advertised. With the enclosed resume, I would like to introduce you to my exceptional planning, organizational, and analytical skills which have been refined in challenging food inspection positions in processing plant environments.

I have completed a number of military and USDA training courses related to food inspection. These included training in the Hazard Analysis Critical Control Point (HAACP) Program, and the Basic and Advanced Food Inspection Preparatory Courses, as well as courses in Workplace Lockout/Tagout Awareness, Basic Livestock Slaughter Inspection, and Introduction to Civil Rights. I am also certified in Community CPR.

As you will see from my resume, I have recently been excelling as Food Inspector for the Food Safety Inspection Service of the USDA, while simultaneously serving as a Medical Specialist in the Army National Guard. During my tenure in this position, I have become very familiar with operational policies and procedures related to food processing, and have worked with facility management and USDA officials in developing HAACP and SSOP policies.

Throughout my career in food inspection, I have demonstrated exceptional attention to detail, and have been cited for my keen observation as well as my for the error-free accuracy of my documentation. My familiarity with food processing operations as well as my ability to evaluate risks and formulate effective preventive measures would make me a strong candidate for the position you advertised.

If you can use a highly skilled professional with exceptional analytical skills and the ability to quickly master and apply complex rules and regulations, I hope you will welcome my call soon when I try to arrange a brief meeting to discuss your goals and how my background might serve your needs. I can provide outstanding references at the appropriate time.

Sincerely,

Martin Daniels

MARTIN DANIELS

1110½ Hay Street, Fayetteville, NC 28305 • preppub@aol.com • (910) 483-6611

OBJECTIVE

To benefit an organization that can use an experienced young professional with exceptional organizational and analytical skills who offers a background in the inspection, documentation, and troubleshooting of products and programs in food processing environments.

EDUCATION

Associate of Science in Biology, Northern State University.
Earned a certificate in **Precision Machining**, Belmont University, Nashville, TN, 1999.
Completed a number of military and USDA training courses related to food inspection, including: Hazard Analysis Critical Control Point (HAACP) Program, 24 hours, 2004; Advanced Food Inspection Preparatory Course, 24 hours, 2003; and Basic Food Inspection, U.S. Academy of Health Sciences, 8 weeks, 2002.

CERTIFICATIONS

Hazard Analysis Critical Control Point (HAACP) certification, 2004.
Certified in Community CPR (Adult, Child, and Infant), renew certification in 2008.

EXPERIENCE

FOOD SAFETY INSPECTOR. Brown Packing, Co., Aberdeen, SD (2005-present). Conduct regular inspections of plant facilities and equipment, raw materials, and product in various stages of processing, observing plant operation to ensure compliance with USDA and federal, state, and local rules, regulations, and guidelines.

- Assist with the development and implementation of the facility's Hazard Analysis and Critical Control Points (HACCP) program.
- Perform pre-operative inspections and monitor operational sanitation in all departments, verifying adherence to the facility's Sanitation Standard Operating Procedures (SSOP).
- Oversee quality control procedures for the facility, evaluating plant logs and verification tests to ensure that the operation strictly adheres to USDA standards.
- Performed net weight checks, formulations, pumping checks, and other tests to ensure compliance with policies, procedures, and regulations of processed meat inspection.
- Observed packaging and labeling of variety meats, chits and casings, etc., ensuring accuracy of labeling data and compliance with regulations and applicable standards.

MEDICAL SPECIALIST. U.S. Army National Guard, Aberdeen, SD (2004-present). Plan and coordinate the issue of equipment and inspection of training facilities to ensure the safety and protection of Guard personnel.

- Assess training environment to determine safety requirements; issue personal protective devices and give medical attention to personnel as the situation requires.

FOOD & SANITATION INSPECTOR. U.S. Army, Fort Jackson, SC (2001-04). Provided a wide range of food safety and inspection services to the Fort Jackson Commissary; worked independently, with little or no supervision.

- Inspected incoming shipments of produce, meats, poultry, and dairy products, checking for proper temperature and monitoring compliance with quality assurance standards.
- Conducted sanitation inspections of the warehouse, retail, and dining facilities on a daily basis, to ensure that storage areas comply with federal and USDA regulations.

AFFILIATIONS

Volunteer extensively with the Special Olympics program; served on the local and Area 7 Manager's Team, recruiting volunteers and coordinating site selection and construction, fundraising, and event planning, as well as training athletes; served on the Rules Committee for the state games.

PERSONAL

Excellent personal and professional references are available upon request.

Petroleum products distribution

Date

Exact Name of Person
Title or Position
Name of Company
Address
City, State, Zip

FUEL SAFETY SUPERVISOR

Dear Exact Name of Person: (or Dear Sir or Madam if answering a blind ad.)

I would appreciate an opportunity to talk with you soon about how I could contribute to your organization through my expertise related to petroleum products and cryogenics.

As you will see from my resume, I am considered one of the petroleum industry's leading fuels systems specialists, and I have supervised personnel and fuels distribution operations throughout the Middle East for the past ten years. Most recently I organized and managed a team of fuels specialists transporting fuel products throughout the Middle East. I have also supervised technical specialists while operating 6,000 gallon aircraft servicing and refueling units.

In one job as a Cryogenics Supervisor, I maintained six storage tanks containing more than 50,000 gallons of liquid oxygen and nitrogen, which included overseeing the receipt, storage, issuance, and inventorying of those products. In another job as a Fuels Storage Operator, I was cited for error-free performance while receiving, storing, transferring, inventorying, and documenting transactions for approximately 34 million gallons of bulk aviation fuels annually.

I have received numerous honors and bonuses and have excelled in state-of-the-art training in my field. You would find me to be a dedicated individual who is known for my safety-conscious attitude and attention to detail. I would be delighted to provide outstanding personal and professional references upon request.

I hope you will write or call me soon to suggest a time when we might meet to discuss your needs and goals and how I might serve them. Thank you in advance for your time.

Sincerely yours,

Peter Whitfield

PETER WHITFIELD

1110½ Hay Street, Fayetteville, NC 28305 • preppub@aol.com • (910) 483-6611

OBJECTIVE

To offer my reputation as a dependable, enthusiastic, and hard-working young professional to an organization that can use my experience in the specialized field of aircraft fueling operations along with my exposure to maintenance, supply, and quality control activities.

EXPERIENCE

Am excelling in the fuel safety field working for Exxon in the Middle East:

2003-present: FUEL SAFETY SUPERVISOR. Managed a team of fuels specialists involved in transporting fuel products throughout the Middle East.

- Evaluated potential team members; made selections; scheduled training for personnel coming into the special team.
- Conducted regular inspections and ensured equipment was properly maintained.
- Contributed to the success of numerous international projects vital to national security.

2000-03: FUELS DISTRIBUTION OPERATOR. Drove and operated 6,000 gallon aircraft servicing and refueling units.

- Personally pumped more than 60 million gallons of aircraft fuel in a 1 1/2-year period — an average of 40 million gallons a year — to several thousand aircraft.
- Used mobile and hydrant equipment to dispense fuels and oil products; maintained equipment; prepared required documentation.

1998-2000: CRYOGENICS SUPERVISOR. Maintained six storage tanks containing more than 50,000 gallons of liquid oxygen and nitrogen: oversaw receipt, storage, issuance, and inventorying.

- Conducted regular odor tests and laboratory analyses to ensure quality of products.
- Maintained accurate and up-to-date records of all preventive maintenance performed on storage tanks, purging units, vacuum pumps, and cosmodyne samplers.

1996-98: FUELS STORAGE OPERATOR. Was cited for error-free performance while receiving, storing, transferring, inventorying, and documenting transactions for approximately 34 million gallons of bulk aviation fuels annually.

1995-96: PREVENTIVE MAINTENANCE SUPERVISOR. Worked closely with vehicle maintenance personnel in order to schedule safety inspections as well as scheduled/ unscheduled maintenance on mobile refueling equipment.

HONORS

Received respected bonuses and certificates for outstanding job performance:

"Employee of the Month" nine times and "Employee of the Quarter" twice in four years
"Pumper of the Month" ten times and "Pumper of the Year" one time

TRAINING

Completed approximately 29 weeks of training: 14 weeks (560 hours) related to fuel operations including air transportable and bulk fuel delivery systems and a four-week lab program and another 16 weeks of leadership, survival, and resistance/interrogation training.

EQUIPMENT EXPERTISE

Am qualified to operate R-5, R-9, R-11, and R-12 Mack and Dodge aircraft refueling vehicles. Operate virtually all Air Force forklifts and many of the Army's fuel servicing units.

PERSONAL

Received special training in graphic arts and printing in high school. Am an extremely reliable, dependable, and responsible individual known for my outgoing personality and ability to get along with everyone I meet.

Aircraft maintenance and HAZMAT

PETER NIMROCK

1110½ Hay Street, Fayetteville, NC 28305

preppub@aol.com (910) 483-6611

HAZARDOUS MATERIALS PROGRAM MANAGER

OBJECTIVE

To benefit an organization that can use a versatile professional with expertise related to hazardous materials program management, maintenance management, and troubleshooting and problem solving.

CLEARANCE

Hold Top Secret security clearance, SSNI

EXPERIENCE

HAZARDOUS MATERIALS PROGRAM MANAGER. Department of Defense, Washington, DC (2001-present). Supervised 30 people in meeting support requirements for 33 C-130E aircraft in a combat airlift squadron which provides worldwide support for the war against terrorism. Was responsible for $900 million in support equipment, and managed a 1,200 line-item bench stock and more than 180 composite tool kits (CTK).

- Managed and controlled 22 complete sets of technical orders over $10 million in test equipment, tools, radios, computers, and vehicles.
- Oversaw the Hazardous Materials (HAZMAT) Program, and on a formal performance evaluation was commended for "completely overhauling the unit's HAZMAT program, making it the best of any squadron in the wing."
- In revamping the HAZMAT Program, also revamped the maintenance Hazardous Communications program into one which exceeded stringent safety and environmental standards, and was commended for "aggressively managing the HAZMAT Program for waste accountability and control."
- Earned an Outstanding rating on the 2001 Environmental Compliance/Assessment Management Program.
- Instituted many changes in the Composite Tools Kit (CTK) area which greatly enhanced productivity; identified and procured $200,000 in badly needed tools for flightline operations.
- Installed a new bar code inventory system to track tool usage which resulted in better accountability and reduced the incidence of lost tools by 50%.
- Was evaluated in writing as a "top performer whose positive attitude motivates his troops and energizes his superiors."

PRODUCTION SUPERINTENDENT. Department of Defense, Classified locations worldwide (1999-00). Extensively involved in Quality Assurance (QA), managed a 40-person work center; provided trained, qualified maintenance technicians for isochronal inspections.

- Managed 45 aircraft maintenance personnel and 10 separate maintenance specialties during ongoing isochronal aircraft inspections as well as an average of four aircraft isochronal inspections monthly.
- Supervised all quality production, safety, security, and foreign object damage prevention requirements in the KC-135 maintenance dock.

- Orchestrated 33 isochronal (ISO) inspections with a quality evaluation rating of 99.7% and zero ISO backlogs.
- Instilled a "safety first and quality assurance first" attitude in all employees which resulted in 99% customer satisfaction; led the organization to win the Commander's Organizational Quality Award.
- Conducted hazardous waste training for shop personnel on proper storage of hazardous material; received zero significant findings from the 2000 Environmental Compliance Assessment Management Program (ECAMP).

INSPECTION DOCK CHIEF. Department of Defense, Belgium (1997-99). Managed and supervised maintenance performed during the phase inspections on 18 E-3A aircraft and associated equipment valued at $3.9 billion. Coordinated work scheduled for 35 people from 11 NATO nations in 17 different occupational specialties.

- Controlled, maintained, and repaired shop equipment valued at $1.3 million.
- On my own initiative, developed a new work flow for the corrosion-control and aircraft-painting portions of the phase inspection.
- Emphasized safety and identified the safety hazards associated with the aircraft dry polish; my diligence resulted in the elimination of an unsafe material.
- Following an air abort of an aircraft, took decisive actions during recovery which resulted in replacement of aborted aircraft within 30 minutes.
- Utilized my technical knowledge and resourcefulness to benefit NATO on numerous occasions: on one occasion, discovered and corrected a fleet-wide discrepancy on nose landing gear lubrication fitting. On another occasion, built an inspection timetable which consolidated the phase and corrosion prevention programs.

MAINTENANCE MANAGER. Department of Defense, Korea (1996-97). Managed a multimillion-dollar flying-hour budget while directing and managing all maintenance performed on aircraft producing an average of 12 sorties and 80 flying hours per month.

- Distinguished myself in all aspects of troubleshooting and quality assurance; once identified the problem causing an intermittent nose-steering problem and saved NATO the cost of contracting a depot team to repair the problem.

Other Department of Defense experience:
AIRCRAFT MAINTENANCE TECHNICIAN. Supervised and accomplished depot-level maintenance inspections and modifications of aircraft systems and components on C/KC-135, B-1B, and B-52 aircraft. Developed a training program for personnel.
SHIFT SUPERVISOR. Supervised test measurement and diagnostic equipment, tool room operations, and shop personnel involved in aircraft maintenance.
Excelled in jobs as a Recovery Team Member, Aircraft Maintenance Specialist, Assistant Crew Chief, Unit Bench Stock Monitor, Non-Powered AGAE Mechanic, and Aircraft Mechanic.

HONORS

Received 28 different awards for distinguished performance.

EDUCATION

Hazardous Waste Operation and Emergency Response (HAZWOPER) Certification Course.
Licensed Aircraft and Powerplant (A&P) Mechanic. FAA Airframe and Powerplant Certified.
Completed training programs related to aircraft maintenance, care and use of test equipment, quality assurance, hazardous waste and environmental awareness, corrosion control and fire prevention, chemical warfare, OSHA and EPA standards, aircraft battle damage repair, leadership and management.
Skilled in using test equipment and diagnostic measurement tools.

SKILLS

Certified Crane operator (15-ton wheel hydraulic); Certified De-icer Truck Basket/Vehicle operator; Certified forklift operator, gas and diesel; Certified Tow vehicle (MB-2, MB-4).

Hazardous waste management

Date

Exact Name of Person
Title or Position
Name of Company
Address
City, State, Zip

HAZARDOUS WASTE MANAGER

Dear Exact Name of Person: (or Dear Sir or Madam if answering a blind ad.)

I would appreciate an opportunity to talk with you soon about how I could contribute to your organization through my knowledge and experience in environmental science, hazardous waste management, safety program development, and a working knowledge of all Federal ACC and AFOSH guidelines and regulations.

As you will see from my resume, I am currently serving my country in the U.S. Air Force as a Hazardous Waste Manager, overseeing 22 satellite waste collection points, developing and coordinating waste management programs, and creating evacuation plans and protective equipment guidelines.

In addition, I have spearheaded my organization's group environmental issues and attended ACC and AFOSH training conferences at the local, state, and Federal levels. I was praised by California state inspectors and U.S. Air Force top management for receiving zero violations during two surprise California Natural Resources Conservation Commission inspections.

You would find me to be a hardworking, dedicated professional who has a reputation for possessing the highest moral standards and integrity. I pride myself on always giving 110% to every job I undertake.

I hope you will call or write me soon to suggest a time convenient for us to meet and discuss your current and future needs and how I might serve them. Thank you in advance for your time.

Sincerely,

Angelica Winters

ANGELICA WINTERS

1110½ Hay Street, Fayetteville, NC 28305 • preppub@aol.com • (910) 483-6611

OBJECTIVE

To obtain an employment/education (COOP) opportunity with a dynamic and growing Environmental Company where I can utilize my military and environmental experience while completing my education in environmental sciences.

EDUCATION

Attended Antelope Valley College, Lancaster, CA, and California State University, San Bernardino, CA.

- Member of Kappa Delta Chi National Honor Society.
- Consistently earned Dean's List honors.

TRAINING

Completed training which included courses emphasizing staff management, written and verbal communication, safety, core automated maintenance systems, cargo rail systems, HAZMAT handling and waste management, ACC Resource Conservation and Recovery Act, computer operations, integrated materials management systems, and environmental issues in programs designed for the military executive; certified to teach ACC environmental issues.

EXPERIENCE

Have built a strong background in diverse areas impacting on public and organizational environmental policies during my 18 years in the U.S. Air Force; am known as a decisive professional offering a reputation for ingenuity and vision:

HAZARDOUS WASTE MANAGER. Edwards AFB, CA (2004-present). Learned the importance of complying with State and Federal Resource and Conservation Recovery Act (RCRA) and Environmental Protection Agency (EPA) laws and regulations while monitoring and analyzing organizational environmental issues and their impact on the local area.

- Directed operations and performed special inspections on 22 satellite collection points and one 90-day collection point.
- Presented lectures to organizational personnel on effective and safe waste management operations, in addition to developing, coordinating, and implementing waste management programs; maintained MSDS catalog. Developed evacuation plans and personnel protective equipment/clothing guidelines. Recognized as the most effective organizational manager for my expertise in controlling a $1 million inventory.
- Singled out to oversee a modernization program, ensured state-of-the-art equipment was integrated into the inventory system, and replaced obsolete equipment.
- Refined both my verbal and written communication skills while developing operations plans concerning waste management and routinely briefing senior executives.
- Lauded by state inspectors and top-level management for receiving zero violations during surprise Natural Resources Conservation Commission inspection.

HAZARDOUS MATERIAL COORDINATOR. Edwards AFB, CA (2002-04). Enforced AFOSH standards for handling and disposing of hazardous waste and ran corrosion prevention program, while also training, supervising, and evaluating 12 personnel.

- Performed special inspections and coordinated the Hearing/Respiratory Protection Program, in addition to developing comprehensive Hazardous Waste Programs.

MAINTENANCE MANAGER and **FIRST-LINE SUPERVISOR.** Edwards AFB, CA (2000-01). Known as an exceptional performer who could be counted on to produce outstanding results; was in charge of $75,000 in aircraft maintenance equipment and parts inventory while also overseeing 20 employees. Handled daily hazardous material inspections and developed a detailed HAZMAT material program for department.

MEMBERSHIP

Member of National Registry of Environmental Professionals, Registration # AEP 6611.

Public health programs

Date

Exact Name of Person
Title or Position
Name of Company
Address
City, State, Zip

HEALTH QUALITY SERVICES SUPERVISOR

Dear Exact Name of Person: (or Dear Sir or Madam if answering a blind ad.)

With the enclosed resume, I would like to make you aware of my interest in exploring employment opportunities with your organization. I am responding to your recent advertisement for an Executive Director.

As you will see from my resume, I have worked for the Shawnee County Health Department since 1985, and I have excelled in a track record of promotion to my current position as Supervisor of the Women's and Children's Health Program. While working within the county's health department, I have become accustomed to interacting with multiple clinics and multiple programs. In my current position, I hire, train, and manage up to 25 individuals while planning and administering multiple budgets totaling more than $2.5 million. I have earned a reputation as a caring individual who is skilled at building consensus and inspiring others to work toward common goals.

An outgoing and energetic individual, I take great pride in the multiple accomplishments of the county's health department, and I have played a key role in many important programs. I co-developed the Shawnee County Healthy Living Program which provided preventive health screening services to the county's 2,000 employees. I have also played a key role in the Pregnant Living Program which has reduced the incidence of teen pregnancies. In addition to organizing numerous projects related to breast cancer awareness and other areas, I developed the Childhood Poison Prevention Program and the Heart Control Program.

While serving the health care needs of the county's indigent population, my main "hobby" has been gaining advanced knowledge through earning additional academic credentials. In addition to earning my L.P.N. and R.N. credentials, I received a Bachelor of Science in Nursing and a Master's in Public Health degree. I am proficient with numerous software programs which I have utilized in my job in order to prepare budgets, track expenditures, and control the funding of multiple programs.

I can provide outstanding personal and professional references at the appropriate time, but I would ask that you not contact the Shawnee County Health Department until after we have a chance to discuss your needs. Since I am in a key management role, I wish my interest in your organization to remain confidential at this time. Thank you in advance for your consideration and professional courtesies.

Yours sincerely,

Lavina Cleveland

LAVINA CLEVELAND

1110½ Hay Street, Fayetteville, NC 28305 • preppub@aol.com • (910) 483-6611

OBJECTIVE

To benefit an organization that can use a strong leader and resourceful problem solver who offers proven experience in motivating employees and inspiring teamwork, managing finances and preparing budgets, as well as in developing and implementing new programs.

COMPUTERS

Proficient with software including Excel, Access, Word, and PowerPoint.

EDUCATION

Master's in Public Health, Emporia State University, Emporia, KS, 2004. GPA 3.8.
Bachelor of Science in Nursing (B.S.N.), Friends University, Wichita, KS, 1999.
Associate's Degree in Nursing (A.D.N.), Wichita State University, Wichita, KS, 1989.
Licensed Practical Nurse (L.P.N.), Washburn University of Topeka, Topeka, KS, 1985.

EXPERIENCE

Shawnee County Health Department, Topeka, KS. Have excelled in the following track record of promotion within the public health field while continuing my education in my spare time. Have worked in a public health environment with multiple clinics serving populations across the life span, from birth to old age.

2000-present: HEALTH QUALITY SERVICES SUPERVISOR. Was promoted in 2000 to supervise the Women's and Children's Health Program, and have earned a reputation as a caring leader who is skilled at inspiring others to work toward common goals.

- **Employee supervision:** Hire, train, and supervise up to 25 individuals who include RNs, LPNs, community health assistants, nurse practitioners, and secretaries. Devise and implement regular inservice training for staff.
- **Budget preparation and management:** Plan and administer multiple budgets totaling more than $2.5 million for activities pertaining to child health, family planning, breast cancer awareness, rape crisis, lead programs, and outreach activities. Create spreadsheets in Excel in order to monitor expenditures and assure that disbursements are within state guidelines.
- **Executive decision making:** Am part of monthly management team meetings, weekly supervisory meetings, and annual Board of Director meetings.
- **Program development:** Played a role in developing and implementing the **Healthy Living Program** which provided 2,000 county employees with preventive health screening and health education. Played a major role in managing the **Pregnant Living Program** funded by the state to decrease unwed pregnancies.

1991-00: CLINIC CHARGE NURSE. Supervised up to three RNs, aides, and communicable disease specialists while working in the Family Health Clinic.

- **Project management:** Organized projects with the Shawnee Regional Hospital to promote health concepts such as breast cancer awareness. Developed the **Childhood Poison Prevention Program** and the **Heart Control Program.**

1989-91: STAFF NURSE. Worked as a School Health Nurse at Birchcreek Junior High and Sunshine Elementary; made home visits and performed physical assessments; rotated as a clinical nurse among all health department clinics including Child Health, Family Planning, TB Chest, Immunization, Maternity, Neurology, Neuromuscular, Orthopedic, and Dental.

1985-89: LPN. Assisted MDs and worked as a Liaison Nurse to Shawnee Regional Hospital, transitioning patients to the from the public health environment; taught classes.

AFFILIATIONS

President, KS Public Health Nurse's Administrators; **Member,** Healthy Choice for Humanity; KS Public Health Association, Shawnee County School Health Committee.

Continuous process flow cells

Date

Exact Name of Person
Title or Position
Name of Company
Address
City, State, Zip

INDUSTRIAL ENGINEER
with experience as a
cost reduction coordinator

Dear Exact Name of Person: (or Dear Sir or Madam if answering a blind ad.)

I would appreciate an opportunity to talk with you soon about how I could contribute to your organization through my industrial engineering background including my experience in managing cost reduction programs, planning capital expenditures, and supporting new product design.

In my current job as an Industrial Engineer and Cost Reduction Coordinator, I have implemented the new manufacturing concept known as continuous process flow cells and have functioned as the "inhouse expert" in training my associates in this area. While managing a $700,000 cost reduction program, I investigate and implement cost reductions through alternative materials and manufacturing processes as well as design modifications. I am involved on a daily basis in on-the-floor problem solving, costing of component processing, tooling and gaging, and capital equipment acquisitions. I have had extensive experience in project management.

Prior to graduating with my B.S. degree in Industrial Engineering, I worked my way through college in jobs in which I was involved in producing computer-aided drawings and participating in new product design. Although I worked my way through college, financing 80% of my education, I excelled academically and received the Outstanding Senior Award.

I am knowledgeable of numerous popular software and drafting packages. I offer a proven ability to rapidly master new software and adapt it for specific purposes and environments.

Single and willing to relocate, I can provide outstanding personal and professional references. I am highly regarded by my current employer, PRP Industries, and have been credited with making numerous contributions to the company through solving problems, cutting costs, determining needed capital equipment, and implementing new processes. I am making this inquiry to your company in confidence because I feel there might be a fit between your needs and my versatile areas of expertise.

I hope you will call or write me soon to suggest a time convenient for us to meet and discuss your current and future needs and how I might serve them. Thank you in advance for you time.

Sincerely yours,

Trent Warner

TRENT WARNER
1110½ Hay Street, Fayetteville, NC 28305 • preppub@aol.com • (910) 483-6611

OBJECTIVE

To add value to an organization that can use an accomplished young industrial engineer who offers specialized know-how in coordinating cost reductions, experience in both manufacturing and process engineering, proven skills in project management, and extensive interaction with product design, quality control, vendor relations, and capital expenditures.

EDUCATION

Bachelor of Science (B.S.) degree, Industrial Engineering Major with a concentration in manufacturing, University of South Dakota, Vermillion, SD, 2002.
- Achieved a 3.5 GPA (3.8 in my major); inducted into Alpha Beta Chi Honorary Fraternity.
- Received **Outstanding Senior Award** in manufacturing concentration.
- Worked throughout college and financed 80% of my education.

Associate of Applied Science (A.A.S.) degree, Mechanical Engineering and Design Technology Major, University of South Dakota, Vermillion, SD, 1998; achieved 3.7 GPA.
From 2002-present, completed continuing education in these areas:

ISO 9000 Internal Auditing
Total Quality Management
Continuous Flow Manufacturing
Root Cause Analysis
Value Engineering/Value Analysis
Synchronous Manufacturing

TECHNICAL KNOWLEDGE

Software: Microsoft Word, Excel, PowerPoint, Adobe
Drafting: VERSACAD, CADAM, Cascade, Intergraph, Unigraphics

CERTIFICATIONS

Certified Manufacturing Technologist; Certified ISO 9000 Internal Auditor

EXPERIENCE

INDUSTRIAL ENGINEER/COST REDUCTION COORDINATOR. PRP Industries, Co., Aberdeen, SD (2004-present). Responsible for the processing of machined components from raw material to finished product while also coordinating a $700,000 annual cost reduction program; investigate and implement cost reductions by exploring the possibility of alternative materials, other manufacturing processes, and design modifications.
- Involved on a daily basis in on-the-floor problem solving, costing of component processing, tooling and gaging, and capital equipment acquisitions.
- Implemented and coordinated continuous process flow cells, a new concept in the manufacturing area; completed extensive training and trained my associates.
- Performed cost justifications and complete equipment installs for capital equipment acquisitions totaling half a million dollars. Continuously interact with new product teams, problem-solving groups, purchasing specialists, vendors, as well as quality control.

ASSOCIATE MANUFACTURING ENGINEER. Dakota Manufacturing, Co., Aberdeen, SD (2003). Coordinated project workloads, designed assembly tooling, and established data bases for tracking and calibration of gaging used in the shop; gained experience related to self-directed work teams, facilities layout, and routing procedures.

Other experience:
DESIGNER. For the Drafting Corporation (1998-02), produced computer-aided drawings and actively participated in new product design while interacting with engineering and manufacturing. Introduced the first microprocessor controlled cruise control.

AFFILIATIONS

Society of Manufacturing Engineers; National Association of Industrial Technology; Alpha Beta Chi International Honorary Fraternity for Education in Technology

PERSONAL

Single; will relocate. Accustomed to hard work and tight deadlines. Have excellent references.

Retail shoplifting prevention

Date

Exact Name of Person
Title or Position
Name of Company
Address
City, State, Zip

LOSS PREVENTION DETECTIVE

Dear Exact Name of Person: (or Dear Sir or Madam if answering a blind ad.)

With the enclosed resume, I would like to make you aware of the knowledge related to finance which I could put to work for you.

While completing my B.S. degree in Finance, I played a key role as a member of a team which analyzed the Harley-Davidson company and made strategic and operational recommendations for the company and industry as a whole. I have earned a reputation as an insightful analyst and problem-solver and I am certain I could make a significant contribution to the bottom line of a company that can use an astute young financier.

In jobs which I held to finance my college degree, I worked in several roles within Macy's Department Store, where I excelled in handling responsibilities as a Loss Prevention Detective, Supervisor, and handler of cash accounting. Although Macy's has strongly encouraged me to remain with the corporation after college graduation and seek internal promotions, I have decided to explore other opportunities.

I can assure you in advance that I have an excellent reputation and would quickly become a valuable asset to your organization. Please contact me if my considerable abilities interest you, and I will gladly make myself available for a personal interview at your convenience.

Sincerely,

David Yuen

DAVID YUEN

1110½ Hay Street, Fayetteville, NC 28305 • preppub@aol.com • (910) 483-6611

OBJECTIVE

To offer my education in finance as well as my analytical, sales, and communication skills to an organization that can benefit from my strong interest in financial planning and banking as well as my personal reputation for integrity, high moral standards, and a strong work ethic.

EDUCATION

Completing a **Bachelor's degree in Finance,** The University of Kentucky, Lexington, KY; degree expected spring 2006.

- Placed on the university's Dean's List in recognition of my academic accomplishments.
- Received an "A" on an intensive class project: performed a company analysis on Harley-Davidson including keeping records, analyzing price and volume data as well as technical data, gathering and analyzing information about the industry, and making determinations on the economic outlook for the company and industry.
- Completed specialized course work such as Finance 330 (principles of finance, stock valuation, options, etc.) and Finance 331 (real estate investing).

EXPERIENCE

Learned to manage time wisely while maintaining at least a 3.4 GPA in my college career and excelling in demanding part-time jobs including this track record of accomplishments with Macy's Department Store, Lexington, KY:

LOSS PREVENTION DETECTIVE. (2003-present). In only 18 months with the company, have progressed to the highest level available to a part-time employee based on my maturity, willingness to take on hard work, and communication skills.

- Increased apprehensions of shoplifters 50%, thereby greatly reducing losses from theft.
- Displayed the ability to remain calm and in control and act as an arbitrator under intense conditions.
- Provided security for the store premises, researched discrepancies in cash accounts, and generated surveillance programs.
- Learned the importance of confidentiality while guarding privileged information.

FRONT-LINE SUPERVISOR. (2002). Supervised approximately 50 employees in order to ensure that customers received the highest quality of service and satisfaction.

- Opened cash drawers and initiated changeovers while register contents were transferred as well as changing large denominations of bills for smaller ones as needed.
- Approved refunds, lay-a-ways, and purchases by associates.
- Conducted new employee orientation which included such areas as cash handling procedures, customer service techniques, and company policy.
- Was honored as **"Associate of the Quarter"** by management and other associates.

CASH OFFICE ASSOCIATE. (2001). Was given the opportunity to apply my knowledge gained in college in a real-life situation while handling day-to-day retail store office activities.

CASHIER. (2000). Became skilled in handling refunds and sales accurately and quickly while becoming responsible for large amounts of cash transactions.

- Was known for my ability to assist customers as well as my keen eye for possible theft.

TRAINING

Completed several seminars and training programs including a Microsoft workshop and loss prevention training (detecting losses, detaining suspects, and making reports).

PERSONAL

Keep up with stock market and read *"The Wall Street Journal"* regularly. Familiar with Windows, and the Internet. Graduated from high school with honors.

Helicopter systems repair

Date

Exact Name of Person
Title or Position
Name of Company
Address
City, State, Zip

MAINTENANCE SAFETY SUPERVISOR

Dear Exact Name of Person: (or Dear Sir or Madam if answering a blind ad.)

I would appreciate an opportunity to talk with you soon about how I could contribute to your company's maintenance excellence through my experience in all areas of maintenance and repair related to the AH-64 Apache helicopter.

As you will see from my resume, I have excelled in jobs as a crew chief, flight line chief, technical inspector, maintenance phase team leader, and floor supervisor. I offer a proven ability to maximize safe flying hours through my technical problem solving abilities as well as through my strong maintenance management skills.

In my most recent job I was named **Employee of the Year** while excelling as a Maintenance Floor Supervisor at the one of the U.S. military bases. I was specially chosen as a technical inspector during the war in Afghanistan and I learned about the unique maintenance requirements of the harsh Afghan climate.

You would find me in person to be an aggressive problem solver who can always be counted on for leadership in "crisis maintenance" situations as well as for the kind of strong time management skills that keep operations flowing efficiently in routine circumstances.

I hope you will call or write soon to suggest a time convenient for us to meet and discuss your current and future needs and how I might serve them. Thank you in advance for your time.

Yours sincerely,

Paul Joseph

PAUL JOSEPH

1110½ Hay Street, Fayetteville, NC 28305 • preppub@aol.com • (910) 483-6611

OBJECTIVE

To benefit an organization that can use a highly skilled aircraft repairman and maintenance manager with exceptional troubleshooting and "crisis maintenance" management skills along with experience as a crew chief, flight line chief, technical inspector, and maintenance shop supervisor.

TRAINING

Excelled in over a year of rigorous technical training in areas including **airborne operations**, **air movement operations**, and **maintenance management**.

- Am certified as a Senior Rated Jumpmaster.

HONOR

In 2005, was named **Employee of the Year** after excelling in a week-long test of leadership.

CLEARANCE

Was entrusted with a Secret security clearance.

EXPERIENCE

MAINTENANCE SAFETY SUPERVISOR. U.S. Army, Fort Riley, KS (2004-present). Supervise and train up to 35 people while applying production control, quality control, and other maintenance management principles to the operations of a shop which diagnoses, troubleshoots, and repairs malfunctions of helicopter subsystems.

- Have increased phase time by over 50% and have boosted the battalion's readiness to the highest level since Saudi Arabia.

TECHNICAL INSPECTOR. U.S. Army, Fort Riley, KS, and Afghanistan (2003-04). Was handpicked for this job during the War on Terror; became familiar with maintenance requirements particular to the harsh Afghan climate while coordinating the outloading of AH-64 Apache helicopters and support equipment.

- Supervised "crisis maintenance" and battle damage repairs while performing expert troubleshooting during combat conditions.
- Became knowledgeable about all areas of logistics pertaining to the deployment of the AH-64 Apache helicopter.

FLIGHT LINE CHIEF. U.S. Army, Fort Benning, GA (2002-03). Supervised up to eight people and managed daily flight schedules and personnel assignments for six aircraft; directed maintenance crews on timely repair assignments to ensure that demanding flight schedules were met.

CREW CHIEF. U.S. Army, Fort Benning, GA (2000-02). After excelling in extensive technical and management training, applied my expert understanding of the Apache organization's procedures while managing up to three people removing and installing subsystem assemblies and components including the following:

engines	gearboxes	transmissions	flight controls
starters	generators	inverters	lights
batteries	pumps	reservoirs	valves
rotors	batteries	hydraulic cylinders	lines

- Prepared helicopters for extensive inspections and maintenance checks by removing items such as cowling, inspection plates, panels, doors, and auxiliary equipment.

PERSONAL

Am considered one of the military's leading experts on the Apache AH-64 helicopter. Am also qualified as an AH-1 Cobra crew chief/mechanic and as an OH-58 Kiowa technical inspector.

Waste removal and disposal

Date

Exact Name of Person
Title or Position
Name of Company
Address
City, State, Zip

MANUFACTURING HAZARDOUS WASTE DIRECTOR & SAFETY CHIEF

Dear Exact Name of Person: (or Dear Sir or Madam if answering a blind ad.)

With the enclosed resume, I would like to make you aware of my background as an experienced, self-motivated, and educated industrial and manufacturing engineer with excellent communication and organizational skills and a background in manufacturing, quality assurance, and safety.

In my most recent position as Hazardous Waste Director and Safety Chief, I was responsible for the security and maintenance of more than $9 million worth of equipment, and developed and managed a $1.5 million budget for the maintenance department. Through proper process and materials selection, I reduced hazardous waste generated by the plant by 99%. I also oversaw the removal and proper disposal of all hazardous materials from the paint line and paint booth after that operation ceased.

In previous positions with Westfield, I was responsible for training employees on environmental, health, and safety issues and personally addressing any problems in these areas. Through my initiative, the facility drastically reduced its generation of hazardous waste, which resulted in a downgrading of our Hazardous Waste Generator Status to Conditionally Exempt.

My innovative ideas were displayed when, as a Senior Manufacturing Engineer, I designed a wiring harness for one of our products, resulting in a $100,000 per year reduction in the manufacturing costs for that item. As Management Systems Supervisor, I managed three departments simultaneously, including the product transfer team that moved nearly $10 million worth of production materials to Chicago from a facility in another state in order to support the addition of a new product to our production line and the expansion of our operation.

If you can use a highly skilled manufacturing engineer with strong communication and problem-solving skills as well as a background which includes hazardous materials handling and safety, I hope you will contact me soon. I can assure you in advance that I have an outstanding reputation and could rapidly become a valuable addition to your company.

Sincerely,

Christopher T. Abraham

CHRISTOPHER T. ABRAHAM

1110½ Hay Street, Fayetteville, NC 28305 • preppub@aol.com • (910) 483-6611

OBJECTIVE

To benefit an organization that can use an experienced, self-motivated, and educated industrial engineer with excellent communication and organizational skills and a background in manufacturing, quality assurance, and safety.

EDUCATION

Master's degree in Engineering, Chicago State University, Chicago, IL, 1997.
Certificate in Business Administration, Keller Graduate School of Management, Chicago, IL, 1982.
Bachelor of Science in Industrial Engineering, University of Illinois, Urbana, IL, 1974.

EXPERIENCE

After a distinguished career with ***Westfield, Inc.****, have taken an early retirement in order to pursue other opportunities; advanced in the following track record of promotion:*

HAZARDOUS WASTE DIRECTOR & SAFETY CHIEF. Chicago, IL (2000-present). Perform a variety of manufacturing, safety, and maintenance functions, taking on additional responsibilities while still performing the duties of Manufacturing Engineer.

- Supervise 14 employees in the maintenance department and tool room; improved on-time performance of all preventative maintenance tasks from 25% to 98%.
- Through proper process and materials selection, lowered the amount of hazardous waste generated drastically, resulting in a downgrade in our Hazardous Waste Generator Status to Conditionally Exempt, reducing future liability and exempting us from certain regulations.
- Reduced hazardous waste generated by 99%.
- Responsible for the security and maintenance of over $9 million worth of equipment.
- Developed and managed a $1.5 million budget for the maintenance department.
- Managed the International Standards Organization (ISO) equipment calibration system.
- Oversaw removal and proper disposal of all hazardous materials from the paint line and paint booth after operation ceased, reducing future liability and returning the area to productive use.

SENIOR MANUFACTURING ENGINEER. Chicago, IL (1994-2000). In addition to the customary duties of a Senior Manufacturing Engineer, I took on additional responsibility for hazardous waste management, safety training and special projects.

- Addressed all problems related to and developed employee training on environmental, health, and safety issues; administered the Worker's Compensation program.
- Reduced hazardous waste generated by the facility 25% and effected compliance with OSHA guidelines and regulations regarding materials handling.
- Oversaw completion of special projects, including replacing the facility's roof and HVAC.

SENIOR MANUFACTURING ENGINEER. Chicago, IL (1990-1994). Directed the troubleshooting and resolution of engineering and manufacturing problems to ensure the smooth operation of the production department

- Trained all new employees in my assigned areas of the facility.
- Developed a wiring harness which reduced production costs by $100,000 per year.
- Directed the preparation of new releases of existing products.

MANAGEMENT SYSTEMS SUPERVISOR. Chicago, IL (1984-1990). Managed three different departments in order to ease the transition to computerization of product engineering; supervised 13 employees in the maintenance and tool room department.

PERSONAL

Outstanding personal and professional references are available upon request.

New product development

Date

Exact Name of Person
Title or Position
Name of Company
Address
City, State, Zip

MECHANICAL ENGINEER
with experience as a director, new product development

Dear Exact Name of Person: (or Dear Sir or Madam if answering a blind ad.)

With the enclosed resume, I would like to make you aware of my interest in exploring the possibility of joining your executive team in some capacity in which you can utilize my vast experience related to new product development and strategic planning/positioning.

As you will see from my resume, I am currently excelling as Director, New Product Development, for Philly Manufacturing Company. I was recruited by the company in 2003 to take over new product development for its Courtney Group and Friller Products ($90 million in sales) and, in May 2004, I was promoted to direct new product development for all company products ($200 million). Although I am held in the highest regard and can provide outstanding references at the appropriate time, I would ask that you not contact my current employer until after we talk. The company I work for is currently up for sale, and I am selectively exploring opportunities in other organizations.

In previous positions since earning my B.S. degree in Mechanical Engineering, I have gained experience in design engineering, process development engineering, machine design engineering, project management, and new product development in multiple industries. I worked for giants (Hershey's, Inc. and Sylvania Corporation) in the aircraft industry and made major contributions to new engine proposals and new aircraft production. Subsequently working on the development of consumer products for Hershey's, I was promoted to develop manufacturing processes for new cereal products. Then with Calvin & Bros., Inc., I was promoted from Senior Project Manager to New Product Development Group Manager. While at Calvin, I transformed a poorly organized group suffering from low output into a highly focused and productive product development team which developed numerous profitable new products.

I offer a proven ability to bring focus and strategic direction to product development teams, and I would welcome the opportunity to meet with you in person to discuss how I might positively impact your bottom line. If you think my considerable skills and experience could benefit you, please contact me to suggest the next step I should take in exploring the possibility of becoming a valuable part of your executive team.

Yours sincerely,

Jeffrey Kinston

JEFFREY KINSTON

1110½ Hay Street, Fayetteville, NC 28305 • preppub@aol.com • (910) 483-6611

OBJECTIVE

To contribute to the growth and profitability of a company that can use a visionary business leader with expertise related to all aspects of new product development including project management, strategic planning, project justification and prioritization as well as employee recruiting and supervision.

EDUCATION

Bachelor of Science (B.S.) degree in Mechanical Engineering, Providence College, Providence, RI, 1989. Graduated *cum laude;* 3.72 GPA.
Annually attend the respected executive development program at St. Joseph's University, Philadelphia, PA, The Masters Forum, 2000-04.

EXPERIENCE

DIRECTOR, NEW PRODUCT DEVELOPMENT. Philly Manufacturing Company, Altoona, PA (2003-present). Was recruited by this company to grow its top line through new product development; was promoted in May 2004 from Director of New Product Development for Courtney Group and Friller Products ($90 million in sales) to Director of New Product Development for all company products ($200 million).

- Oversee a $5 million engineering budget and a staff of 38 managers, engineers, designers, drafters, technicians, and assistants; report to the President.
- Closed down new product development operations in Virginia and rebuilt the new product development organization in Pennsylvania; hired 18 managers, engineers, designers, and technicians while also organizing lab facilities and test equipment.
- Played a key role in identifying and arresting faltering financial performance; developed and implemented a disciplined new product development process that included organizing and flow-charting the process, developing work instructions, creating process and authorization forms, and designing a financial model for control.
- Developed and launched a new hydraulic jack, a retail 12-volt portable power supply, a gasoline recycler, and a cordless device utilized in manufacturing.

GROUP MANAGER, NEW PRODUCT DEVELOPMENT. Calvin & Bros., Inc., Pittsburgh, PA (1998-03). Excelled as a Senior Project Manager from 1998-03 and was promoted in 1999 to New Product Development Group Manager; reported to the VP of Engineering.

- Facilitated the design and development of high quality innovative industrial fluid-handling products at the lowest cost and in the shortest possible time.

PROCESS DEVELOPMENT ENGINEER. Hershey's, Inc., Pittsburgh, PA (1995-98). After making major contributions as a Machine Design Project Engineer from 1995-97, was promoted to develop manufacturing processes for new snack products.

- Served as the Process Engineering Representative on new product development teams.
- Defined, tested, and installed new processing equipment; developed and implemented procedures needed to transition new products from R&D labs to plant production.

STRUCTURAL INTEGRITY ENGINEER. Sylvania Corporation, Yakima, WA (1994-95). Received two Sylvania Achievement Awards for contributions to the F120 engine proposal.

DESIGN ENGINEERING LEAD. American Designers, Inc., Richmond, VA (1989-94). Excelled as a Structural Design Engineer from 1989-92 and designed primary aircraft structure for new production F-16 aircraft, and was then promoted to manage all structural design and production support related to the F-16 aft fuselage. Supervised eight engineers.

PERSONAL

Excellent references. Highly resourceful leader who excels in managing the creative process.

Decontamination and operation

Date

Exact Name of Person
Title or Position
Name of Company
Address
City, State, Zip

NUCLEAR SAFETY TRAINING MANAGER

Dear Exact Name of Person: (or Dear Sir or Madam if answering a blind ad.)

Can you use an experienced hazardous waste inspector and operations manager who offers a "track record" of approximately 12 years of experience in nuclear/biological/chemical (NBC) operations?

While working in the nuclear field, I have become known as a "subject matter expert" on all aspects of inspection procedures, new product fielding, and training. In my current position with the Nuclear Safety Agency as a Nuclear Safety Training Manager, I am involved in overseeing nuclear training programs for multiple organizations.

As you will see from my resume, my experience includes service during the War on Terror in the key role of setting up and running a continuous training program for 3,200 people. Setting up this program and then putting it into action in an actual combat situation resulted in important information which I used to streamline NBC use plans for the future. In recognition of my expertise, I was retained for an extended period from 1999 to 2003 as the "technical expert" for a 12,000-person military base. I was certified as a Chemical Inspector and conducted regular inspections of an $8 million inventory, ensured in-depth training, developed inspection procedures checklists still in use, and dealt with my counterparts throughout the world for the purpose of establishing inspection procedures.

I have been exposed to Nuclear Regulatory Commission (NRC) rules and regulations and have been involved in handling/storing/transporting hazardous materials and feel that I offer a broad base of knowledge and experience which could make me a valuable asset to your organization.

I hope you will welcome my call soon to arrange a brief meeting at your convenience to discuss your current and future needs and how I might serve them. Thank you in advance for your time.

Sincerely yours,

Christopher Simpson

Alternate last paragraph:
I hope you will call or write soon to suggest a time convenient for us to meet and discuss your current and future needs and how I might serve them. Thank you in advance for your time.

CHRISTOPHER SIMPSON

1110½ Hay Street, Fayetteville, NC 28305 • preppub@aol.com • (910) 483-6611

OBJECTIVE

To apply my managerial, motivational, and organizational abilities to an organization that can use a seasoned professional offering specialized expertise in the inspection, handling, and storage of hazardous waste materials and expertise in training/developing personnel.

EXPERIENCE

NUCLEAR SAFETY TRAINING MANAGER. Nuclear Power Agency, Washington, DC (2005-present). Ensure the nuclear/biological/chemical (NBC) defense readiness of three 700-person organizations and their personnel including training, maintenance, and logistics.

- Managed a $250,000 inventory of 200 items of equipment and 700 protective masks per unit with no discrepancies noted in numerous inspections.
- Coordinated training when the new M40/42 protective mask was issued.
- Developed standard operating procedures (SOP) for marshaling personnel during Emergency Deployment Readiness Exercises (EDREs).

COMBAT NBC OPERATIONS MANAGER. U.S. Army, Iraq (2004-05). As the "subject matter expert" on NBC operations, developed/organized/managed a program of continuous training for 3,200 personnel which directly impacted on the safety of the organization's personnel during the War on Terror. Applied information gained during actual combat to develop a streamlined utilization plan for NBC and its equipment to be used in the future.

INSPECTION TEAM TECHNICAL EXPERT. U.S. Army, Fort Campbell, KY (1999-03). Retained well beyond the normal period of service because of my recognized expertise, served as the point-of-contact on chemical related policies/regulations for a 12,000-person organization and the liaison with Army units worldwide for establishing inspection standards.

- Conducted regular inspections of $8 million worth of NBC equipment/protective masks.
- Developed the checklists still in use by the 101st Airborne Division for conducting chemical inspections and for preparing chemical defense equipment reports.

CHEMICAL DECONTAMINATION SPECIALIST. U.S. Army, Fort Campbell, KY (1994-99). Supervised an eight-person team which set up and operated decontamination units for 20 to 100 people and five to 20 vehicles at a time.

- Was singled out to find clean water and report on foreign chemical weapons.

EDUCATION & TRAINING

Completed 2 1/2 years of college course work, Central Michigan University.
Excelled in more than 1,260 hours of specialized training programs emphasizing chemical operations, leadership, personnel management, and inspection procedures.

SPECIAL SKILLS & EQUIPMENT EXPERTISE

Operate and maintain NBC equipment including the following:
Nuclear: AN/PDR 27 radiacmeter
IM 174 radiacmeter
IM 93 dosimeter
PP/1578 charger
Chemical: M8 and M8A1 chemical agent detectors
M17 series protective mask
M40/42 series protective mask
M41 Protective Mask Fit Validation System (PMFVS)
Chemical Agent Monitor (CAM)
Decontamination: M12A1 power-driven decontamination unit
M17A3 series decontamination unit
Smoke devices: M3A1 and M3A3 smoke generators

Chemical and radiation detection

Date

Exact Name of Person
Title or Position
Name of Company
Address
City, State, Zip

NUCLEAR SUPERVISOR AND SAFETY OFFICER

Dear Exact Name of Person (or Dear Sir or Madam if answering a blind ad):

I would like to take this opportunity to make you aware of my experience and knowledge in the highly specialized area of environmental field operations and chemical protective services gained while serving my country in the U.S. Army.

As you will see from my enclosed resume, I offer special abilities related to setting up decontamination sites and in conducting chemical reconnaissance surveys and sampling. Through my training and experience, I have also become skilled in troubleshooting and repairing equipment which includes chemical agent monitors, radiation detectors, and chemical agent alarms. I am familiar with Level A protective suits, SCBA (Self Contained Breathing Apparatus), and the M-40 protective mask. My military training emphasized leadership development as well as the technical aspects of safety and quality control; identification, detection, and sampling of industrial hazards; and decontamination, detection, and equipment maintenance. I was the honor graduate of my Chemical Operations Specialist Course.

Widely recognized as a knowledgeable environmental field technician with a talent for training others and providing sound management for resources, I consistently brought about improvements to the units where I was assigned. In my most recent job at Fort Campbell, KY, the nation's busiest military base worldwide, I developed accountability and maintenance programs where none had previously existed. I was credited with taking "the worst NBC Room" in the parent organization and turning it into one which set the standard for other similar programs in an 18,000-person organization.

In a prior assignment in Italy, I received training from the PRP Corporation and was sent to Iraq to sample, detect, and identify industrial hazards. I became trained on the $2 million Mobile Mass Spectrometer used to analyze chemical agents. In my first assignment at Fort Drum in 2001-02, I received an Army Achievement Medal for totally revamping a substandard NBC program.

If you can use a reliable young professional with a reputation for total integrity and honesty combined with strong technical expertise in chemical protective services and environmental field operations, I hope you will call me soon to suggest a time when we might have a brief discussion of how I could contribute to your organization. I will provide excellent professional and personal references at the appropriate time.

Sincerely,

Vincent Edwards

VINCENT EDWARDS

1110½ Hay Street, Fayetteville, NC 28305 • preppub@aol.com • (910) 483-6611

OBJECTIVE

To offer experience and knowledge in the specialized area of environmental field operations and chemical protective services to an organization that can use a strong leader who performs effectively under pressure and enjoys meeting challenges head on.

EDUCATION & TRAINING

Completed courses in **Criminal Justice and U.S. History**, Hopkinsville Community College and Central Texas College, Hopkinsville, KY, campuses.

Graduated from Bloomington High School, Bloomington, IN, 2000.

- Was in ROTC; played baseball.

Received extensive U.S. Army-sponsored training in leadership development and advanced lifesaving as well as in the following technical courses:

- NBC (nuclear/biological/chemical) Recon — reconnaissance, survey, and detection
- techniques and Mobile Mass Spectrometry
- Industrial Hazards – identification, detection, and sampling
- Chemical Operations Specialist Course – basic decontamination, detection, and equipment maintenance
- Safety and Quality Control

AREAS OF EXPERTISE

Offer special abilities in setting up decontamination sites, conducting chemical reconnaissance and surveys, and troubleshooting and repairing equipment including:

Chemical Agent Monitor	AN/UDR-2 and AN/PDR-75 Radiation Detectors
M-8A1 Chemical Agent Alarm	M-40 Protective Mask
Level A Protective Suit	SCBA (Self Contained Breathing Apparatus)

EXPERIENCE

Became known as a knowledgeable environmental field technician, trainer, and resource manager while serving in the U.S. Army:

NUCLEAR SUPERVISOR AND SAFETY OFFICER. Fort Campbell, KY (2005-present). Applied exceptional levels of knowledge and skills while thoroughly revitalizing and revamping NBC support services for an 82-person organization.

- Implemented and conducted training in all aspects of NBC (nuclear/biological/chemical) equipment and individual protective measures and in OSHA regulations and guidelines.
- Maintained and controlled an inventory which included 27 chemical detectors, nine radiation detectors, 90 protective masks and 120 suits, and a small warehouse of other NBC items with a total value in excess of $250,000.
- Was credited with significant improvements which brought the unit "NBC Room" from the worst in the parent organization to second best in the entire 18,000-person division with a 97% rating in one major inspection.
- Turned a "substandard" maintenance program into a top-notch program after creating accountability and maintenance systems where none had previously existed.

NBC OPERATIONS SPECIALIST. Italy (2001-04). Gained extensive and thorough knowledge with an emphasis on all aspect of chemical surveys, reconnaissance, and detection as operator of a $2 million vehicle-mounted computer system used to analyze chemical agents, the Mobile Mass Spectrometer.

- Was selected to receive training by the PRP Corporation for a mission in Iraq: was trained to sample, detect, and identify industrial hazards; learned to use the level A protective suit; and was also trained to use various types of civilian equipment.

PERSONAL

Received several certificates and letters of achievement and commendation including recognition as honor graduate of my NBC course and for outstanding results in inspections.

Petroleum and fuel handling

Date

Exact Name of Person
Title or Position
Name of Company
Address
City, State, Zip

PETROLEUM SUPPLY AND ISSUE SPECIALIST

Dear Exact Name of Person: (or Dear Sir or Madam if answering a blind ad.)

I would appreciate an opportunity to talk with you soon about how I could contribute to your organization through my specialist knowledge of petroleum supply receipt, storage, and issue.

While serving my country in the U.S. Army, I earned a reputation for my technical proficiency and gained experience in a variety of facilities and locations throughout the world. I was handpicked to spend a year in Bosnia with the Multinational Force and Observers and then spent two months in Kuwait during preparations for the conflict in Kuwait. One of my most important contributions in Bosnia came when I prevented a 250,000 fuel spill which resulted in saving the government more than $140,000 in lost inventory.

My experience has included supervising teams of specialists, issuing aircraft fuel, controlling documentation of fuel transactions, and ensuring quality of facilities and procedures.

I hope you will welcome my call soon to arrange a brief meeting at your convenience to discuss your current and future needs and how I might serve them. Thank you in advance for your time.

Sincerely yours,

Alexis Blackwell

Alternate last paragraph:
I hope you will call or write soon to suggest a time convenient for us to meet and discuss your current and future needs and how I might serve them. Thank you in advance for your time.

ALEXIS BLACKWELL

1110½ Hay Street, Fayetteville, NC 28305 • preppub@aol.com • (910) 483-6611

OBJECTIVE

To contribute to an organization through my outstanding motivational and communication abilities and specialist knowledge of petroleum supply operations as well as through my initiative, energy, and dedication to producing efficient, high-quality services.

EXPERIENCE

PETROLEUM SUPPLY AND ISSUE SPECIALIST. U.S. Army, Fort Hood, TX (2005-present). Provided an aviation company with fuel products and was recognized with a certificate of achievement for being "instrumental" in allowing my company to successfully complete its assigned responsibilities.

- Issued more than 200,000 gallons of fuel during a task force's special training project.

RETAIL FUEL SERVICES MANAGER. U.S. Army, Bosnia (2004). Was handpicked to provide support for an 11-member multinational peacekeeping force by controlling the receipt, storage, and issue of an inventory of more than one million gallons of diesel fuel and mogas.

- Applied my expertise while preventing a spill of more than 250,000 gallons of fuel, thereby saving the loss of more than $140,000.
- Earned a commendation medal for my professionalism in maintaining 100% accountability for an inventory valued at $70,000 in a 13-line gas cylinder and 17-line 55-gallon drum fuel yard dispensing more than 150,000 gallons monthly.
- Was officially evaluated as a "meticulous professional with an eye for details."

PETROLEUM SPECIALIST. U.S. Army, Kuwait (2003). Controlled the shipment, issue, storage, and receipt of fuels including J-P4, mogas, and diesel to supported units preparing for the conflict in Kuwait.

- Gained experience in working and communicating with others as part of a team.

Earned promotion in this "track record," U.S. Army, Fort Stewart, GA:

SUPERVISORY PETROLEUM SPECIALIST. (2000-03). Directed the performance of seven employees involved in issuing, transporting, and receiving fuel.

- Learned to keep detailed and accurate records of fuel received and issued.
- Was awarded an achievement medal for "professionalism and technical knowledge" while providing support during "Operation Enduring Freedom" in Afghanistan.

AIRCRAFT FUEL SPECIALIST. (2000). Was selected for a three-month special assignment at Hunter Army Airfield and learned the special methods of issuing aircraft fuel.

- Became familiar with different types of military aircraft and how to use hand signals to guide planes into place for refueling.

FUEL MANAGEMENT SPECIALIST. (1999-00). Gained knowledge related to issuing fuel from retail points to supported units and became familiar with proper methods of recording fuel transactions. Earned recognition for "outstanding service" during a major training exercise in South America.

TRAINING

Excelled in leadership and petroleum specialist training programs.

SPECIAL KNOWLEDGE

Offer skills which include operating forklifts along with specialist knowledge of aircraft refueling procedures.
Completed qualification and received the Driver's Badge and the Mechanic's Badge.

PERSONAL

Known for initiative and dedication. Work well independently or as a member of a team.

Disaster relief and community services

Date

Exact Name of Person
Title or Position
Name of Company
Address
City, State, Zip

PROJECT DIRECTOR, DISASTER RELIEF PROGRAM

Dear Exact Name of Person: (or Dear Sir or Madam if answering a blind ad.)

With the enclosed resume, I would like to make you aware of my interest in exploring employment opportunities with your organization.

As you will see from my resume, I enjoy helping others and finding ways to improve the quality of their lives. I am adept at planning large-scale humanitarian relief projects and skilled at organizing the details so that projects are successful and productive.

Currently as Project Director for the Disaster Relief Program, I lead and direct the efforts of a management team which plans and coordinates emergency and long-term response following natural disasters. Part of my job is to plan and supervise the construction of housing and program sites supporting 12,500 volunteers, and I oversee activities which provide food and housing for hundreds of volunteers and staff members. I am a co-developer of successful operational plans in which large groups of volunteers have been used to provide large-scale emergency relief efforts worldwide.

In previous employment, I was a Program Coordinator for a "Migrant Education" program involving more than 170 children in 18 migrant labor camps. In that capacity, I addressed the basic issues of ensuring that children received basic educational service while addressing the spiritual needs of children and families.

I hope you will call or write me soon to suggest a time convenient for us to meet to discuss your current and future needs. Thank you in advance for your time.

Sincerely yours,

Ryan Johnston

Alternate last paragraph:

I hope you will welcome my call soon to arrange a brief meeting when we might meet to discuss your needs and goals and how my background might serve them. I can provide outstanding references at the appropriate time.

RYAN JOHNSTON

1110½ Hay Street, Fayetteville, NC 28305 • preppub@aol.com • (910) 483-6611

OBJECTIVE

To offer my broad base of experience in project management, problem solving, marketing, and administration to an organization that can use my compassion and true concern for others as well as my specialized education and experience in disaster relief and aid projects.

EXPERIENCE

PROJECT DIRECTOR. Disaster Relief Program, Washington, DC (2002-present). Lead and direct the efforts of a management team which plans and coordinates emergency and long-term response following natural disasters.

- Plan and supervise the construction of housing and program sites in support of more than 12,500 volunteers providing relief efforts in countries all over the world.
- Oversee activities which provide food and housing for an average of 350 volunteers and staff members on a daily basis.
- Plan for and manage construction activities including scheduling, logistics support, and purchasing for 280 to 300 construction projects annually.
- Direct multimillion-dollar budgets.
- Played a key role in the development and founding of Northfield Schools, an alternative program for delinquent adolescents.
- Was a co-developer of successful operational plans in which large groups of volunteers were used to provide large-scale emergency relief efforts.

PROGRAM COORDINATOR. Migrant Workers Advocacy Agency, Malletts Bay, VT (1999-01). Helped establish a "Migrant Education" program at Malletts Bay, VT: this was a program involving more than 170 children in 18 migrant labor camps on Malletts Bay. Addressed the basic issues of ensuring that children received basic educational services and, in a nondenominational fashion, also addressed the spiritual issues of children.

PRESIDENT & GENERAL MANAGER. Vermont Builders Supply, Burlington, VT (1999). Established and managed a successful general contracting business which built homes, remodeled upscale homes, and performed light commercial work.

Highlights of earlier experience: Built a unique business which transported large yachts cross country after accomplishing continued growth as a partner in a general residential and light commercial construction company.

- Became the youngest assistant manager and manager for two different hospitality companies. Gained experience in all aspects of the industry from custodial duties, to bartending, to kitchen cook, to catering, to accounting, to reservations.

EDUCATION

Received certification as a trainer following advanced training by the Department of Psychiatry's post-disaster mental health course, University of Vermont, Burlington, VT, 2003.

Completed two years of college course work in Business and Liberal Arts, Champlain College, Burlington, VT.

AFFILIATIONS

Served on the board of directors of organizations which included the following:
State of Vermont Migrant Task Force, 1998-2004
Coalition Board of Directors, 2002-2003

PERSONAL

Enjoy knowing I am helping others and finding ways to improve the quality of their life. Offer a reputation as a compassionate, caring person. Am adept at large-scale planning and organizing the details so that projects are successful and productive.

Personal care products

Date

Exact Name of Person
Title or Position
Name of Company
Address
City, State, Zip

QUALITY ASSURANCE MANAGER

Dear Exact Name of Person: (or Dear Sir or Madam if answering a blind ad.)

I would like to make you aware of my interest in a Quality Assurance position within a corporation which can utilize my strong executive skills as well as my proven ability to apply my technical expertise in resourceful ways that improve the bottom line while strengthening customer satisfaction.

As you will see from my resume, I have worked for the past 17 years for National Industry Company, where I have been promoted to increasing levels of responsibility. In my current job, I manage a 19-person QA department in a plant which employs 425 people and manufactures personal care products totaling $250 million. While managing a departmental budget of nearly $1 million, I have transitioned the plant from a regular production assembly line operation into a team-managed operation in which teams of employees are responsible for individual products. This has shifted QA from a "police" role to a consulting and monitoring role. I have also developed, implemented and managed a Cost-of-Quality Program which achieved 2004 cost savings of $150,000 by identifying and eliminating unnecessary processes.

In my previous job at Quality Assurance Manager at the company's Massachusetts plant, I developed and implemented a Quality Demerit System which the company now uses corporate-wide. The Quality Demerit System transformed the four National Industry manufacturing plants from a quality level of 67% defect-free product to the consumer in 1999, to 98.2% defect-free in 2000. The targeted goal for 2005 is 99.0% defect-free product.

I offer extensive expertise related to blow molding and injection molding. Both in my current job and in my job in Massachusetts, I managed Quality Assurance related to blow molding and injection molding. In the Massachusetts plant, we achieved a 10% improvement in lots accepted when using Mil. Std. 105E to determine acceptable quality levels (AQLs).

You would find me in person to be a congenial individual who prides myself on my ability to establish and maintain effective working relationships. I can provide outstanding personal and professional references at the appropriate time. I hope you will contact me to suggest a time when we might meet to discuss your needs as well as my skills, experience, and qualifications.

Sincerely,

Raymond Peterson

RAYMOND PETERSON

1110½ Hay Street, Fayetteville, NC 28305 • preppub@aol.com • (910) 483-6611

OBJECTIVE

To benefit an organization that can use a knowledgeable quality assurance executive who offers a track record of achievements based on my ability to lower costs, improve customer satisfaction, reduce defects, and strengthen employee accountability.

EXPERIENCE

1988-present: Have excelled in the following track record of promotion with National Industry Company, a $1 billion a year Fortune 500 company which recently became a part of American Industry USA, a $4 billion corporation which also includes Biolage and Dial Soap.

QUALITY ASSURANCE MANAGER. Brookline, MA (2000-present). Manage a 19-person department which includes two QA Supervisors and a Laboratory Manager in a plant which employs 425 people and manufactures personal care products totaling $250 million.

- Oversee quality assurance throughout the plant which includes the monitoring of the blow molding and injection molding operations, off-line silk screening and pressure sensitive labeling, and the packaging of finished products.
- Developed, implemented, and managed a Cost-of-Quality Program which identified and eliminated non-value added process activities; this produced a 2004 saving of $150,000.
- Implemented analytical and microbiological testing of raw materials and bulk products which instituted QA at the earliest possible point in the manufacturing process.
- Through aggressive education and skill training, transformed my Quality Assurance Department into the one widely acknowledged as the best in the corporation.
- Manage a departmental budget which in 2004 was slightly less than $1 million; developed and implemented initiatives which stimulated cost efficiencies, and improved worker participation in all facets of QA.
- Transitioned this plant from a regular production assembly line operation into a team-managed operation in which teams are responsible for individual products. This has shifted QA from a "police" role to a consulting, coordinating, and monitoring role.
- Implemented Employee Information Boards for each team which shows vital information including the top five defects, wastage, and efficiency measures. Developed flexible work schedules which improved customer service and which strengthened customer satisfaction.

QUALITY ASSURANCE MANAGER. Boston, MA (1998-00). In a plant that employed 600 people, managed and directed plant quality efforts through process evaluations, quality measurements, and cycle time reductions; managed ten people and a budget of $275 million.

- Developed and implemented a Quality Demerit System which the company now uses corporate-wide; the Quality Demerit System measure the quality performance of our business units as well as the outcoming quality of line teams.
- The Quality Demerit System transformed the four American Industry manufacturing plants from a quality level of 67% defect-free product to the consumer in 1982, to 98.2% defect-free in 2000. The targeted goal for 2005 is 99.0% defect-free product.

Other experience:

QUALITY ASSURANCE MANAGER. Watertown, MA (1996-98). Managed and directed 50 companies who were contracted to product $200 million in cosmetics, fragrances, and other personal care products for American Industry and Perry Ellis.

INSPECTION SUPERVISOR. Peabody, MA (1988-96). Develope and managed the vendor certification program with the result that the customer service level was raised to 99%.

EDUCATION & TRAINING

Completing **MBA** in my spare time at Harvard University; degree anticipated 2006.
B.A. with Major in **Psychology** and Minor in **Sociology**, Boston University, MA, 1991.
Completed executive development courses including extensive training in QA ISO 9002.

Date

Exact Name of Person
Title or Position
Name of Company
Address
City, State, Zip

QUALITY ASSURANCE MANAGER

Dear Exact Name of Person: (or Dear Sir or Madam if answering a blind ad.)

With the enclosed resume, I would like to make you aware of my interest in exploring employment opportunities with your organization.

As you will see from my resume, I have excelled as a Maintenance Supervisor and Quality Assurance Manager with the U.S. Navy, and I have been described as a "meticulous manager" and "a master at motivating and molding subordinates." Considered an expert at troubleshooting and repairing aircraft systems, I am a skilled troubleshooter in correcting stubborn problems affecting powerplant systems. At locations worldwide, I have trained and supervised junior technicians as well as mid-managers performing maintenance on a fleet of aircraft. During the War on Terror, I was specially selected as Troubleshooting Manager in charge of assuring that aircraft were ready to fly on time in a combat environment.

I am skilled in managing maintenance in work center environments. In my previous job as a Quality Assurance Manager in charge of managing aircraft maintenance managers involved in quality assurance activities, I performed testing for maintenance personnel and conducted audit inspections. I was repeatedly sought out by my peers when they needed help in solving complex and perplexing troubleshooting problems. A resourceful individual with a strong bottom-line orientation, I developed an innovative and time-efficient compass swing procedure which was evaluated for incorporation into SH-60H technical manuals.

If you can use an astute industrial manager who is accustomed to working in an environment in which there is "no room for error," I hope you will contact me to suggest a time when we might meet to discuss your needs. I can provide outstanding personal and professional references at the appropriate time.

Sincerely,

Eric Knight

ERIC KNIGHT

1110½ Hay Street, Fayetteville, NC 28305 • preppub@aol.com • (910) 483-6611

OBJECTIVE

To benefit an organization that can use a talented problem solver who offers extensive experience in maintenance management and quality assurance along with a reputation as a superior leader and supervisor with an ability to train, motivate, and develop others.

SKILLS

Am a skilled aviation electrician, and can expertly troubleshoot/repair electrical and electronic aircraft systems and can read AC/DC circuits on A7-E, FA-18, SA-60B, and 53B aircraft.

- Skilled in cable repair and instrument repair.
- MC-100 Operator; proficient with C compasses; FA-18 radar, weapons systems.
- Skilled in soldering, operating oscilloscopes, troubleshooting of RF Iff and tower amplifiers. Proficient with solid state power supplies.
- Licensed to operate numerous vehicles including tow tractors, fork lifts, and buses.

EXPERIENCE

QUALITY ASSURANCE MANAGER. U.S. Navy, Norfolk, VA (2005-present). Was handpicked for this position which involves managing individuals involved in performing maintenance on a fleet of aircraft; prioritize workload, schedule work to be done, oversee ordering of spare parts and repair parts, and continuously train personnel in effective troubleshooting techniques.

- On a technical performance evaluation, was commended for "great personal initiative in assuring that the highest standards of safety and quality were always maintained."

DIVISION SUPERVISOR & QUALITY ASSURANCE MANAGER. U.S. Navy, Pensacola, FL (2001-04). Was evaluated as an "absolute top performer" while managing 11 senior aircraft maintenance managers involved in quality assurance activities. Conducted evaluations of work centers under my supervision, and performed testing for maintenance personnel. Was described in writing as "a master at motivating and molding subordinates."

- Was described as a "meticulous manager" after producing top-notch results in directing 15 comprehensive quarterly audits; maintained highest standards of quality and safety.
- With a reputation as a superb technician, was sought out by my peers and supervisors when they needed help solving complex and perplexing troubleshooting problems.
- Developed an innovative and time-efficient compass swing procedure, which was evaluated for incorporation into SH-60H technical manuals.

AVIONICS-ELECTRONICS SUPERVISOR. U.S. Navy, New Orleans, LA (1996-01). While supervising 15 individuals, received a respected Navy Commendation Medal and was evaluated as "the single most important petty officer in this command" and a "masterful technician and manager who demonstrates the highest level of integrity and judgment."

- Was cited for unparalleled knowledge of the F/A-18 electrical/avionics and weapons systems. Was cross trained in all areas of F/A-18 maintenance and quality assurance.
- On numerous occasions, diagnosed non-routine and complex malfunctions which threatened to interfere with high-tempo flight operations. Supervised comprehensive aircraft inspections; interpreted electrical and electronic schematics and drawings.

TRAINING

Extensive U.S. Navy training in management and areas including safety and quality:

- Navy "A" School (380 hours): Electricity and Electronics
- Aviation Electricity: AC/DC Circuits and vacuum tubes
- F-18 Electrical Instrumentation and organ national maintenance (240 hours)
- Quality Assurance and Safety (numerous courses and hours)

PERSONAL

Reputation as a dynamic leader known for integrity and absolute reliability.

Antenna and communications systems

Date

Exact Name of Person
Title or Position
Name of Company
Address
City, State, Zip

QUALITY ASSURANCE SUPERINTENDENT

Dear Exact Name of Person: (or Dear Sir or Madam if answering a blind ad.)

With the enclosed resume, I would like to make you aware of my distinguished background of effectiveness in managing communications systems installation projects and of my strong knowledge of such diverse areas as technical school operations, procurement, and inventory control.

As you will see from my resume, I built a reputation as an adaptable professional while excelling in managing multiple simultaneous projects in demanding technical environments for the U.S. Air Force. Accustomed to working under harsh weather conditions and within tight deadlines, I have overseen projects in the Arctic, South America; and throughout Europe.

Presently pursuing an associate's degree after earlier studying Electronics Systems Technology, I also received extensive military training which emphasized management and leadership as well as techniques for effective instructors. In my final military assignment prior to retirement, I served as Chief of Communications Systems Construction with the responsibility for planning, coordinating, and managing projects throughout the world. As operations director of a 55-person work center, I oversaw antenna and communications systems installation, enhances, and reconstruction projects which ensured continuous communications support for activities vital to national security.

Earlier jobs in quality assurance inspection, training program instruction and management, and engineering installation operations supervision allowed me to refine my reputation as a creative, energetic professional who could be counted on to complete projects on time and within budget. Cited on numerous occasions for solving difficult problems in ways which reduced expenditures and still allowed the project to be completed ahead of schedule, I was often sought out to provide technical guidance as well as counseling for subordinates.

If you can use an experienced communications technician with excellent time and resource management skills, I hope you will welcome my call soon when I try to arrange a brief meeting to discuss your goals and how my background might serve your needs. I can provide outstanding references at the appropriate time.

Sincerely,

Troy Hanson

Alternate last paragraph:

I hope you will write or call me soon to suggest a time when we might meet to discuss your needs and goals and how my background might serve them. I can provide outstanding references at the appropriate time.

TROY HANSON

1110½ Hay Street, Fayetteville, NC 28305 • preppub@aol.com • (910) 483-6611

OBJECTIVE

To offer a strong background in the management of communications systems installation and operations to an organization that can benefit from my leadership and expertise.

EDUCATION & TRAINING

Currently attending Motlow State Community College, Tullahoma, TN.
Studied Electronics Systems Technology at the Community College of the Air Force.
Received extensive U.S. Air Force training in management techniques and leadership skills.

TECHNICAL SKILLS & KNOWLEDGE

Communications: use all cable locators and troubleshooting test equipment
Computers: am proficient in Windows XP, Word, Excel, and PowerPoint
Construction equipment: operate trenchers, backhoes, line trucks, and tractor-trailers

EXPERIENCE

Advanced to manage worldwide communications systems, U.S. Air Force:

QUALITY ASSURANCE SUPERINTENDENT. Arnold AFB, TN (2005-present). Provide oversight to a team of 55 communications systems construction professionals, including 11 junior managers and 44 technicians; direct installation, enhancement, and reconstruction of antenna and communications systems in locations throughout the world.

- Plan, coordinate, and oversee multiple simultaneous projects, including installation of a $9 million air traffic control system for ten FAA communications sites and an antenna installation project which saved $35,000 by using salvaged materials.
- Control a $3 million annual operating budget, handling inventory control, purchasing, and procurement; developed a preventive maintenance program for antenna systems.
- "Expertly planned" the removal of a major antenna system and its relocation without damage to structural steel, avoiding $50,000 in expenses.
- Locate and correct a design error in eight National Weather Service Next Generation Radar systems, reducing their vulnerability to lightning strikes.

DIRECTOR OF PLANS AND OPERATIONS. Arnold AFB, TN (2003-04). Managed a technical school program, supervising a staff of three personnel providing instruction to over 300 students; developed lesson plans and coordinated training sites and materials.

QUALITY ASSURANCE INSPECTOR. Arnold AFB, TN (2002). Supervised two quality control instructors while managing quality initiatives and training programs emphasizing time management, customer service, and human relations; was cited as a "superb evaluator, exceptional facilitator and team leader." Established a new training center.

ENGINEERING INSTALLATION SUPERVISOR. Arnold AFB, TN (2001). Provided on-site supervision and safety guidance during antenna, steel tower, rigid radome, and associated systems installation, removal, and maintenance activities; earned promotion to QA Inspector based on results obtained in this position.

CENTRALIZED ANTENNA SYSTEMS SUPERVISOR. Aviano AB, Italy (1997-01). Managed 20 technicians maintaining $19.5 million worth of antenna systems at 85 NATO sites throughout Europe; oversaw scheduling of installation, repair, and maintenance.

Highlights of earlier experience: Advanced in supervisory roles while applying technical expertise as a communications electronics technician and manager at worldwide locations.

PERSONAL

Was entrusted with a Secret/NATO Secret security clearance. Am technically knowledgeable with a reputation for creativity and a high level of energy and enthusiasm.

Nuclear attack systems

Date

Exact Name of Person
Title or Position
Name of Company
Address
City, State, Zip

QUALITY ASSURANCE SUPERVISOR AND INSPECTOR

Dear Exact Name of Person: (or Dear Sir or Madam if answering a blind ad.)

With the enclosed resume, I would like to make you aware of my background in quality control and supervision as well as my desire to utilize my skills for the benefit of your organization.

As you will see, I am qualified as a Quality Assurance Supervisor and Inspector, and my experience in quality assurance spans more than a decade. Qualified as an Explosive Handling Expert, I am very knowledgeable of all procedures related to hazardous materials handling and transportation. In my most recent position, I provided oversight for ten people involved in day-to-day systems operations and maintenance on a nuclear attack submarine. Through my strong leadership and personal initiative, I transformed a new torpedo handling section into one recognized as "the best" of 20.

In a prior position, I maintained complete budget accountability of more than $250,000 and I have supervised as many as 40 people. I am qualified to operate and maintain both nuclear and non-nuclear systems, and I have trained numerous other individuals to do the same.

You would find me in person to be a versatile individual who is known for attention to detail and unquestioned integrity, as well as exceptionally strong problem-solving and troubleshooting skills. I have become accustomed to working in environments in which there is often "no room for error," and I believe strongly in maintaining the highest emphasis on safety and quality assurance at all times.

If you can use a seasoned quality assurance professional with proven supervisory and management skills, I hope you will contact me to suggest a time when we might meet to discuss your needs. I am available for relocation and travel as your needs require.

Sincerely,

Jeffrey Stevens

JEFFREY STEVENS

1110½ Hay Street, Fayetteville, NC 28305 • preppub@aol.com • (910) 483-6611

OBJECTIVE

To offer my experience in supervising and inspecting quality assurance and control activities as well as my extensive technical knowledge and mechanical skills to an organization that can use an articulate and mature professional known for initiative, energy, and drive.

TRAINING

Extensive U.S. Navy-sponsored training and certified as a **Quality Assurance Inspector**.

- Am qualified as an **Explosive Handling Expert**.

Excelled in programs emphasizing nuclear and non-nuclear weapons loading and handling, submarine repair parts ordering and supply management, as well as security/law enforcement and weapons training.

HONORS

Received **numerous medals and honors** which included two Meritorious Unit Citations, three Navy Achievement, one National Defense, and one Southwest Asia Medals, one Navy and one Armed Forces Expeditionary Medal, and three Sea Service Ribbons.
Nominated for **"Sailor of the Year"** twice after being voted **"Sailor of the Quarter"** twice.
Received eleven **Letters of Commendation** from senior officers in recognition of my professionalism, outstanding service, positive and enthusiastic outlook, and drive.

EXPERIENCE

Built a track record of exceptional performance and was consistently evaluated as a dynamic top-notch performer with superior abilities, U.S. Navy:

QUALITY ASSURANCE SUPERVISOR AND INSPECTOR. San Diego, CA (2005-present). Provide oversight for ten people performing day-to-day system operations on a nuclear attack submarine – systems included air, hydraulic, weapons delivery, refrigeration, oxygen, and nitrogen.

- Received numerous citations for my ability to motivate, lead, and train personnel who were frequently older than me and to guide them to outstanding results.
- Excelled in handling pressure, participating in five successful classified operations, one of which led to the arrest of a major drug trafficker and seizure of $30 million in cocaine.
- Molded a new torpedo handling section into one recognized as **"the best"** of 20.

MILITARY POLICE OFFICER. Key West, FL (2001-04). Ensured physical security and law enforcement for a naval weapons station with almost 11,000 acres of land and 2.5 miles of waterfront; conducted investigations of all types of criminal cases.

- Led an eight-person team carrying out security exercises which tested key restricted areas as well as working with a Navy SEAL team during waterborne training exercises.
- Provided control for the supply department to include conceiving, researching, and developing budget analysis and tracking systems used in managing more than $250,000.

SUPERVISOR. Pearl Harbor, HI (1997-01). Through initiative, enthusiasm for learning, and a positive attitude, advanced from one of the most junior personnel on board a nuclear attack submarine to become one of its most valuable employees holding responsibility for multimillion-dollar equipment and **Top Secret** material.

- Selected to attend several advanced mechanical training schools, was class leader and graduated at the top of each class: learned to read blueprints, system diagrams, electrical schematics, and make repairs to nuclear and non-nuclear systems as well as receiving training in all aspects of mechanics from repairing high pressure air systems to major hydraulic systems with some training in refrigeration and oxygen system repair.

PERSONAL

Am a team player who contributed to the efforts of intramural football and softball teams which won recognition for their winning records. Was entrusted with a Top Secret clearance.

Parachute rigging

Date

Exact Name of Person
Title or Position
Name of Company
Address
City, State, Zip

QUALITY CONTROL ADVISOR and TECHNICAL SUPERVISOR

Dear Exact Name of Person: (or Dear Sir or Madam if answering a blind ad.)

With the enclosed resume, I would like to make you aware of my versatile background in employee supervision and training, quality control inspection, and supply as well as specialized technical operations.

As you will see from my resume, I am serving in the U.S. Army where I have earned numerous medals in recognition of my technical expertise, accomplishments, and skills. While gaining a reputation as an enthusiastic and energetic professional, I have advanced to supervisory roles ahead of my peers and been selected for numerous special projects in international settings.

Presently assigned to a job where quality control inspections are a major part of my duties, I also am heavily involved in training new personnel in the technical aspects of safely rigging parachutes for U.S. Special Forces personnel. Prior to this assignment, I served in Italy where I participated in support roles in operations, training, supply, and employee supervision. I coordinated all aspects of the training and preparation of personnel for airborne operations in a 115-person organization with facilities for landing, equipment, and aircraft support. During this period I earned Army Commendation and Achievement Medals as well as a Joint Service Achievement Medal for my professionalism and technical expertise in planning for and packing parachutes and other equipment and supplies for numerous training exercises and real-world missions.

My earliest military assignment at Aberdeen Proving Ground, MD, allowed me opportunities to contribute during the testing phase of the research and development process for new airborne extraction and air drop systems for personnel and equipment applications. I was awarded several medals which consistently cited my "meticulous attention to detail" and "extraordinary efforts and accomplishments."

If you can use an experienced supervisor who is known for his high levels of drive, initiative, and energy, I hope you will welcome my call soon when I try to arrange a brief meeting to discuss your goals and how my background might serve your needs. I can provide outstanding references at the appropriate time.

Sincerely,

Patrick Hayes

PATRICK HAYES

1110½ Hay Street, Fayetteville, NC 28305 • preppub@aol.com • (910) 483-6611

OBJECTIVE

To contribute through a versatile background which has included supervising and training others and performing technical quality control inspections to an organization that can benefit from my positive approach, dedication, and energetic and enthusiastic attitude.

TRAINING

Completed extensive military training with an emphasis on leadership development and employee supervision as well as the technical areas of airborne operations, parachute rigging (honor student), and hazardous material transportation and control.

SPECIAL SKILLS

Vehicle operations: am licensed and experienced in the operation of up to 10,000-lb. forklifts, tractor trailers, overhead cranes up to 40 tons, and up to 5-ton vehicles

Security clearance: have been entrusted with a Secret security clearance

EXPERIENCE

Recognized as a detail-oriented professional who can be counted on to improve productivity and increase levels of team accomplishment, U.S. Army, 1998-present:

QUALITY CONTROL ADVISOR and **TECHNICAL SUPERVISOR.** Fort Richardson, AK (2005-present). As a Parachute Rigger for a psychological operations unit which responds to situations worldwide, supervise as many as four people while also packing personnel parachutes and ensuring that all equipment is prepared according to regulations.

- Emphasize attention to detail and safety while training new personnel.
- Observe parachute operations and if any malfunctions occur, advise supervisors to stop the mission so technical inspections can take place and problems be corrected.

OPERATIONS SUPERVISOR and **TRAINING SPECIALIST.** Italy (2002-05). After excelling as a Parachute Rigger, was promoted to this supervisory role in 2004 to coordinate the details of arranging for training, aircraft support, facilities, and equipment for 115 people in order for them to polish skills and maintain current jump status.

- Gained experience in supply support, preparing documentation and arranging for the shipment of hazardous materials through a Special Forces transportation site.
- Was awarded an Army Commendation Medal for accomplishments including repacking more than 40 parachutes with no malfunctions for a special training exercise in Italy.
- Received an Army Achievement Medal for "meritorious service" in recognition of leadership skills and technical expertise displayed while training personnel: this allowed a large-scale international training exercise which was carried out "flawlessly."
- Earned a respected Joint Service Achievement Medal for "distinguished service" during the 2003 deployment to Iraq for Operation Iraqi Freedom; provided leadership for the movement of 200 people and 750,000 pounds of equipment and supplies into Iraq.

QUALITY TESTING SUPERVISOR. Aberdeen Proving Ground, MD (1998-02). Advanced as a Parachute Rigger and gained recognition for my supervisory skills while overseeing special projects and managing as many as 15 civilian and military personnel.

- Contributed to the successful completion of research and development activities for parachute extraction systems and different types of air drop and airborne systems.
- Earned my first Army Commendation Medal and two Army Achievement Medals.
- Credited with "demonstrating technical competence and untiring efforts" which were vital to the success of the rigging and air drop of 500 test bundles of various sizes and configurations which were used to support the air drop phase of a humanitarian mission.
- Participated in freefall operations from 25,000 feet to test a special resupply system.

PERSONAL

Have basic language skills in Italy. Am widely known for my drive and energy.

Aviation systems

Date

Exact Name of Person
Title or Position
Name of Company
Address
City, State, Zip

QUALITY CONTROL INSPECTOR & SAFETY CHIEF

Dear Exact Name of Person: (or Dear Sir or Madam if answering a blind ad):

With the enclosed resume, I would like to make you aware of my interest in exploring employment opportunities within your organization.

As you will see from my enclosed resume, I gained versatile knowledge and experience in numerous areas while serving my country with distinction in the U.S. Air Force. Originally trained as an Aircraft Mechanic, I was subsequently trained in hazardous waste management, quality assurance, safety, and automated inventory control. I was selected for leadership and management training and, upon graduation from Airman Leadership School, I received the award given to the graduate who was "most outstanding in academic testing, performance evaluations, and demonstrated leadership as determined by student and faculty votes."

In my most recent position, I was singled out as playing a major role in helping the Air Force Base win prestigious awards for safety excellence. As Training Manager, I supervised training for 125 crew chiefs and oversaw up to 40 people involved in performing aircraft maintenance.

In a previous assignment, I was rapidly promoted to Assistant Shop Chief in charge of four maintenance personnel while handling numerous other responsibilities. As hazardous waste monitor, I provided oversight for the safe handling and disposition of chemicals and other materials. While in California, the Air Force base converted from a fleet of 126 F-111s to a fleet of 60 F-16s. I coordinated the sale of F-111 parts and equipment to the British Royal Air Force while setting up a whole new shop and ordering a multimillion-dollar inventory for the new fleet. In a prior job at Andersen AFB, Guam, I played a key role in that base's receiving its first "excellent" maintenance rating in over five years.

I am proficient in using automated systems for inventory ordering and control, and I have become skilled at obtaining hard-to-locate parts in an era when supplies were scarce.

I hope you will contact me to suggest a time when we might meet in person to discuss your needs. I can provide excellent references.

Sincerely,

David Constance

DAVID CONSTANCE

1110½ Hay Street, Fayetteville, NC 28305 • preppub@aol.com • (910) 483-6611

OBJECTIVE

To benefit an organization that can use a background which includes extensive technical training, personnel training and supervision, operations management, and inventory control.

EDUCATION

College: Nearly two years of college courses, Columbus Community College, 1994-95.
Technical: Completed **Aircraft Maintenance and Repair Course**, Greensboro, NC, 1998.

- Subsequently completed Hazardous Waste Management Training which covered fire and safety procedures, handling and transportation, record keeping, and planning.
- Training in The Quality Concept, including process evaluation and improvement.
- Received more than twelve Certificates of Training from schools covering subjects including the F-15E Engine Run Qualification Course and aircraft maintenance training.

Leadership: Completed Airman Leadership School, Beale AFB, CA, 2002.

- Received PRP Award presented to the individual in the graduating class who is *"most outstanding in academic testing, performance evaluations, and demonstrated leadership as determined by student and faculty votes."*

Computers: Utilize core database and automated inventory systems; Excel; Word.

EXPERIENCE

Was promoted ahead of my peers because of my management ability and problem-solving skills and frequently worked in a position usually reserved for a higher-ranking professional, U.S. Air Force, locations worldwide.

QUALITY CONTROL INSPECTOR & SAFETY CHIEF. Columbus AFB, MS (2004-present). Played a key role in this Air Force base's winning the American Safety Award for maintenance excellence as well as the highest maintenance award given annually by the Air Force.

- Because of my technical knowledge, received a special waiver to function at the 7 skill level although I was formally trained at the 5 skill level.
- Through my management ability and relentless attention to detail, played a key role in achieving outstanding results on Operation Enduring Freedom.
- As Training Manager, supervise training for 125 crew chiefs. Supervise up to 40 people involved in aircraft maintenance; train aircraft technicians in safety and quality control.

SAFETY COORDINATOR, ASSISTANT SHOP CHIEF, & TRAINING MANAGER. Beale AFB, CA (2000-03). Began as a Repair and Reclamation Specialist and was promoted to Assistant Shop Chief; supervised four people and was in charge of ordering, receiving, and controlling a multimillion-dollar inventory of equipment and supplies which supported a fleet of 126 F-111s.

- Was Training Manager for a 600-person squadron supervising the scheduling of training.
- When the unit converted from F-111s to F-16s, coordinated the equipment support part of the transition; started up a whole new shop for F-16s; ordered and controlled a multimillion-dollar inventory; coordinated the sale of F-111 parts to British Air Force.
- Acted as hazardous waste monitor and oversaw the safe handling and disposition of grease chemicals, cleaning fluids, byproducts and other materials.
- Provided oversight for hazardous waste storage according to OSHA and EPA standards.

MAINTENANCE TECHNICIAN & NUCLEAR SURETY INSPECTOR. Andersen AFB, Guam (1998-00). Played a key role in this base receiving its first "excellent" rating in over five years; was personally singled out for my expert management of the CORE automated maintenance system, a database of every maintenance activity performed on the base.

PERSONAL

Excellent personal and professional references on request. Highly motivated, reliable person.

Logistics and supply

Date

Exact Name of Person
Title or Position
Name of Company
Address
City, State, Zip

QUALITY CONTROL INSPECTOR

Dear Exact Name of Person: (or Dear Sir or Madam if answering a blind ad.)

I would appreciate an opportunity to talk with you soon about how I could apply my attention to detail, energy, and dedication to excellence for the benefit of your organization.

While serving my country in the U.S. Air Force I became extremely knowledgeable and experienced in the field of logistics with an emphasis on quality control inspection/evaluation procedures and inventory control management.

Handpicked for my most recent job as a Quality Control Inspector, I was officially described as having "a perfect record of outstanding performance and clear leadership potential." Chosen for this special role as "policeman" for a chief executive, my job was to know the technical aspects of supply management, review supply programs, make recommendations, and ensure compliance with proper procedures.

In earlier roles including Inventory Control Manager, Requisitioning Technician and Specialist, and Material Control Specialist, I advanced in knowledge to positions of greater responsibility handling multimillion-dollar inventories, special programs, and record-keeping actions while also finding ways to decrease errors and improve efficiency.

I offer a reputation as a quick learner who works well with others as a team member as well as in supervisory roles. I am also widely recognized as a "hard charger" who will not quit until the job is completed to perfection.

I hope you will welcome my call soon to arrange a brief meeting at your convenience to discuss your current and future needs and how I might serve them. Thank you in advance for your time.

Sincerely yours,

Julia Williams

Alternate last paragraph:

I hope you will call or write soon to suggest a time convenient for us to meet and discuss your current and future needs and how I might serve them. Thank you in advance for your time.

JULIA WILLIAMS

1110½ Hay Street, Fayetteville, NC 28305 • preppub@aol.com • (910) 483-6611

OBJECTIVE

To contribute through my outstanding knowledge of inventory control management and quality control inspection procedures to an organization that can use a detail-oriented, energetic, and resourceful young professional.

EXPERIENCE

QUALITY CONTROL INSPECTOR. U.S. Air Force, Ramstein AB, Germany (2005-present). Handpicked by a chief executive to conduct technical inspections, oversee supply operations for 61,000 line items of supplies and equipment valued at more than $351 million while "troubleshooting" potential problem areas.

- Was involved in the purchase of $17.5 million in supplies in addition to $38 million worth of jet fuel needed to support 16,000 flights a year.
- Gained the respect of senior personnel by always remaining "accurate and constructive" when delivering inspection results and making suggestions.
- Reorganized a system of 1,700 records in 26 separate filing systems.
- Review inspection procedures and developed an in-depth checklist.

Advanced in the following "track record," U.S. Air Force, Davis-Monthan AFB, AZ:

2004-05: INVENTORY CONTROL MANAGER. Reduced excess property by $26,000 to below the 3% standard level for the first time in the organization's history while managing a 2,382-item inventory and a redistribution program.

- Located a source which provided more than $4,400 in free equipment which allowed crucial repairs to be made despite delays in normal channels.

2003-04: ADMINISTRATIVE ASSISTANT. Maintained perfect accountability of 277,076 parts in 8,800 locations; implemented regular inventories on 19,728 line items valued at $4,125,408; researched and documented discrepancies.

- Controlled 170 equipment accounts valued at almost $50 million.
- Processed more than $181,000 worth of "special use" transactions.

2002-03: LOCAL PURCHASE REQUISITIONING TECHNICIAN. Polished my technical computer operating skills to input information on prices, quantities, adjustments, cancellations, questions, and status upgrades.

- Was singled out to aid another unit during a severe personnel shortage.
- Managed the implementation of a specialized automated system.

2001-02: REQUISITIONING PROCEDURES SPECIALIST. Controlled programs which accounted for and redistributed excess equipment; monitored supply requests; followed through to see that requests were met in a timely manner.

- Recovered thousands of dollars worth of usable equipment by preparing quarterly brochures and weekly messages listing available items.

2000-01: SUPPLY DOCUMENTATION CLERK. Controlled the receipt, processing, and distribution of supply records for approximately 150 work centers.

- Introduced procedures which reduced errors and increased productivity.

TRAINING & EDUCATION

Excelled in extensive training programs including an advanced internship for inventory management specialists which awarded 45 credit hours toward a degree in Logistics Management from the Community College of the Air Force.

COMPUTER SKILLS

Offer computer operations experience and am familiar with software including Adobe PageMaker and Microsoft programs.

PERSONAL

Earned a commendation medal for "outstanding professional skills, knowledge, and tireless efforts." Speak and read German. Top Secret security clearance.

Electrical safety

Date

Exact Name of Person
Title or Position
Name of Company
Address
City, State, Zip

QUALITY CONTROL INSPECTOR

Dear Exact Name of Person: (or Dear Sir or Madam if answering a blind ad.)

I would appreciate an opportunity to talk with you soon about how I could contribute to your organization through my excellent technical electronics skills, expertise in troubleshooting and fault isolation, and supervisory abilities.

You will see from my resume that I offer a strong background of accomplishments. With eight years of experience with synthetic aperture systems and five with infrared and laser targeting, I am current with state-of-the-art electronics.

I enjoy a challenge and feel that I could make important contributions in new system development projects through the combination of my "hands-on" technical skills and quality control experience. My working knowledge extends to microwave test and diagnostic equipment as well as the standard electronics test equipment and I have good soldering skills.

My communication skills, both verbal and written, have been described as "precise and informational." I have contributed my knowledge while rewriting technical orders and procedures and have written operational instructions and maintenance awareness bulletins which have been accepted and adopted for use worldwide.

I hope you will welcome my call soon to arrange a brief meeting at your convenience to discuss your current and future needs and how I might serve them. Thank you in advance for your time.

Sincerely yours,

Martin W. Zikorsky

Alternate last paragraph:
I hope you will call or write me soon to suggest a time convenient for us to meet and discuss your current and future needs and how I might serve them. Thank you in advance for your time.

MARTIN W. ZIKORSKY

1110½ Hay Street, Fayetteville, NC 28305 • preppub@aol.com • (910) 483-6611

OBJECTIVE

To apply my excellent technical electronics skills and troubleshooting abilities for an organization that can use my experience in maintenance, production, and quality control.

TECHNICAL EXPERTISE

- Troubleshoot and repair to the component level, radar and targeting sensors on the Low Altitude Navigation and Targeting Infrared for Night (LANTIRN) system manufactured by Martin Marietta Electronic Systems, the Advanced Synthetic Aperture Radar System (ASARS2) by Hughes Aircraft, and the CAPRE radar system by Goodyear Aerospace.
- Operate test equipment including: Hewlett Packard 54111D digitizing oscilloscope, 438A average power meter, and 70000 spectrum analyzers; Tektronix 1240 logic analyzer; Racal Dana universal counter; Fluke multimeters; Wavetek 8501 peak power meter.
- Program the DEC PDP 11/44 computer in FORTRAN, Assembler, and Basic languages.
- Can troubleshoot the DEC Microvax II to the circuit card level.
- Offer technical writing experience which has included rewriting technical orders and procedures and writing maintenance operating instructions and maintenance awareness bulletins.

EDUCATION

A.A.S., Avionics, Community College of the Air Force, 1988.

TRAINING

Excelled in more than 800 hours of technical and leadership training including:

LANTIRN advanced maintenance — 190 hours
ASARS2 advanced maintenance — 172 hours
leadership and human relations — 160 hours
advanced management and leadership — 240 hours

EXPERIENCE

QUALITY CONTROL INSPECTOR. General Dynamics, Los Angeles, CA (2000-present). Evaluated as an "exceptional" performer and "outstanding technician," was selected for the critical job of inspecting avionics maintenance personnel and recommended corrective actions.

- Played a "vital role" on an F-100 jet engine inspection which resulted in the development of findings which significantly reduced engine downtime and aircraft abort rates.
- Developed a more effective system of reporting defective parts.
- Prevented LANTIRN pod access panel damage by correcting faulty procedures.
- Wrote an electrical safety guidelines bulletin for maintenance personnel.

PRODUCTION SUPERVISOR. General Dynamics, Los Angeles, CA (1998-00). Was consistently sought out for my "technical expertise" and "exceptionally effective leadership" while involved in prioritizing work flow and directing/controlling repairs in support of the LANTIRN systems for F-15 and F-16 aircraft.

- Brought about a 20% production increase and a "complete turnaround" in morale.
- Developed a method for recovering a coolant which resulted in first-year savings of $32,800 and eliminated environmental hazards.

Highlights of U.S. Air Force career: Served as a **PRODUCTION SUPERVISOR** (1995-97) in maintenance operations; **SUPERVISORY RADAR SYSTEMS SPECIALIST** (1992-94), and **AVIONICS TECHNICIAN** (1988-94).

PERSONAL

Earned numerous awards and medals including three commendation medals and the Southwest Asia Service Medal for service in support of the war in the Middle East.

Industrial machines

Date

Exact Name of Person
Title or Position
Name of Company
Address
City, State, Zip

QUALITY CONTROL INSPECTOR

Dear Exact Name of Person: (or Dear Sir or Madam if answering a blind ad.)

With the enclosed resume, I would like to make you aware of my interest in exploring employment opportunities with your organization.

As you will see from my resume, I have worked as a CNC Technician and Machinist for companies including McDonnell Douglas, Channel Machines, and C & C Industrial Machines. Accustomed to working with extremely close tolerances, I am experienced in working with a wide range of materials including titanium, aluminum, magnesium, lead, and astroline. I am proficient in setting up CNC vertical lathes and mills.

Prior to relocating to Bangor, I worked as a CNC Set-Up Technician for a job shop which completed contracts for companies including Boeing, Harris Semiconductor, and Symmetric. While setting up machines in a job shop, I gained experience with G&M programming, and I worked with vertical milling machines. Prior as a CNC Milling Set-Up Specialist for McDonnell Douglas, I was involved in setting up vertical machines for the production of aircraft parts.

Most recently, I worked as a CNC Machinist and Quality Control Inspector for a company which made parts for DuPont. As a Quality Control Inspector, I walked the production floor to inspect the quality output of 16 different machines, and I also trained other CNC operators and worked as a CNC Machinist.

I have established an outstanding safety record and can provide outstanding references at the appropriate time. I hope I will have an opportunity to meet with you in person to discuss your need for a highly qualified CNC Machinist or quality control professional.

Yours sincerely,

Ethan Manchester

ETHAN MANCHESTER

1110½ Hay Street, Fayetteville, NC 28305 • preppub@aol.com • (910) 483-6611

OBJECTIVE

To benefit an organization that can use a CNC machinist who offers an extensive quality control background along with the set up and production experience of working with materials including the following: titanium, aluminum, magnesium, and lead.

EXPERIENCE

QUALITY CONTROL INSPECTOR. Channel Machine, Bangor, ME (2005-present). For a company which makes parts for DuPont, handle responsibilities as a CNC Machinist and Quality Control Inspector. Walk the floor to provide quality assurance for 16 different machines; make adjustments as necessary. Operate a CMM in the Quality Control Lab, and set up a Brown & Sharp CMM and a Zeise CMM to check parts. Train employees in quality control techniques, which include training in the use of flower gages, micrometers and veniers to check parts.

CNC TECHNICIAN & SET-UP TECHNICIAN. C & C Industrial Machines, Augusta, ME (2002-05). Began as a CNC Operator and advanced to Set-Up Technician with this company which maintained a job shop environment. The company's machines were frequently changed as C & C completed different contracts for customer companies including Boeing and Symmetrics.

- Became skilled in handling hazardous materials while performing as a CNC Machinist; worked with materials which included magnesium, titanium, aluminum, and lead.
- Set up vertical mills, CNC lathes, and CNC screw machines.
- Became skilled in setting up machines in a job shop. Learned about G&M code programming. Cut jaws; made first-piece inspections and adjusted offsets to compensate for parts dimensions. Used CAM system on a Mazak machine to program parts. Used C & C's vertical milling machines and set up vertical milling machines.

CNC MILLING SET-UP SPECIALIST. McDonnell Douglas, Bangor, ME (2000-02). Set up vertical milling machines for the production of aircraft parts. Worked with materials including titanium and aluminum. Became accustomed to working with extremely close tolerances. Became highly proficient with milling set-up techniques.

MACHINIST APPRENTICE. Vester's Machines, Augusta, ME (1995-00). Completed a four-year Machinist Apprenticeship while working for a company which served customers such as BP Oil; was involved in performing duties such as reconditioning valves for BP oil lines.

EDUCATION

Completed four-year apprenticeship program as a **Machinist**, University of Maine at Augusta, ME, 1995-99.
Completed studies in **Political Science** during one year and six months of training with University of Maine at Augusta, ME, 1994-95.
Graduated from Kennebec River High School, Augusta, ME, 1991.
Completed numerous training programs related to safety and machine operation sponsored by McDonnell Douglas and other companies.

SKILLS

Blueprint reading; shop safety; proficient in grinding tools, sharpening drills, and machine setup techniques; skilled in using measuring devices; proficient in shop math. Skilled in working with horizontal and vertical milling machines. Experienced with multiple vice set-ups and fixture set-ups. Skilled with 5 axisis of machining. Proficient with horizontal lathes.
Programs: Familiar with AutoCAD, Fanuc Controls, Siemens Controls and Programming.

PERSONAL

Outstanding quality control skills. Expert knowledge of machine operations.

Tactical vehicle testing

Date

Exact Name of Person
Title or Position
Name of Company
Address
City, State, Zip

QUALITY CONTROL INSPECTOR

Dear Exact Name of Person: (or Dear Sir or Madam if answering a blind ad.)

With the enclosed resume, I would like to make you aware of my background as an experienced industrial equipment specialist who offers a track record of success in operations management and training related to maintenance, repair, and operations of heavy equipment.

In my most recent position as an Instructor and Quality Control Inspector for Vimax International, I was responsible for conducting in-service inspections to ensure that equipment delivered to the Army was safe and fully operational. I trained civilian and military personnel in the performance of troubleshooting to the component level, repair and replacement of defective parts and systems, and maintenance of new Family of Medium Tactical Vehicles (FMTV) equipment. As a Manufacturing Equipment Specialist, I instructed employees in the proper operation and maintenance of the full range of ground support equipment.

With a Bachelor of Science in Business Management from Central Wyoming Community College as well as a two-year diploma in Manufacturing Maintenance Technology, I have a strong educational background to support my years of practical experience.

If you can use an accomplished management professional whose knowledge of industrial equipment maintenance and repair and exceptional training abilities have been tested in challenging environments, then I look forward to hearing from you soon. I assure you in advance that I have an excellent reputation and would quickly become an asset to your organization.

Sincerely,

Samuel D. Dylan

SAMUEL D. DYLAN

1110½ Hay Street, Fayetteville, NC 28305 • preppub@aol.com • (910) 483-6611

OBJECTIVE

To benefit an organization that can use an experienced manufacturing equipment specialist and instructor with strong communication and motivational skills who offers experience in managing all phases of heavy equipment maintenance, repair, and operations.

EDUCATION

Bachelor of Science in **Business Management**, Central Wyoming Community College, Elkhart, WY.
Completed a two-year diploma program in **Manufacturing Maintenance Technology**, Central Wyoming Community College, Elkhart, WY.
Completed a tractor-trailer operator course, Fort Bragg, NC.

LICENSES

Hold a Class "A" Commercial Driver's License (CDL).

EXPERIENCE

QUALITY CONTROL INSPECTOR. Vimax International, Fort Bragg, NC (2000-present). Because of my vast knowledge of automotive manufacturing, was specially selected for this position which involved inspecting the new Family of Medium Tactical Vehicles (FMTV) equipment and road-testing all FMTVs to ensure that the vehicles being delivered to the Army were safe and fully operational before they were put into service.

- Conducted a complete inspection of all vehicles, checking for vehicle damage, missing parts, and proper fluid levels as well as safety and correct operation of all moving parts.
- Performed troubleshooting to the component level to determine the cause of any system malfunctions and repaired or replaced defective parts before issuing the vehicle.
- Served as Assistant Instructor for the FMTV Unit Maintenance and Direct Support level Maintenance Courses.
- Trained military personnel in troubleshooting, maintenance, and repair of electrical, hydraulic, and pneumatic systems, as well as the engine and drive train of FMTVs.
- Provided operator training to military personnel for all truck models, with additional training on the dump truck, cargo crane, and wrecker.

MANUFACTURING EQUIPMENT SPECIALIST. ATCOM, St. Louis, MO (1987-1999). While working for a government contractor, traveled worldwide to provide new equipment training to mechanics and equipment operators in supported units as well as technical, logistical, and supply assistance.

- Attended numerous New Equipment Instructor Training courses in order to effectively coach maintenance and operations personnel on appropriate procedures.
- Conducted Component Testing and Repair classes for Direct Support Level Maintenance Mechanics.
- Trained personnel on the operation and maintenance of various types of equipment to support the ground operation of troops, which included:

Generators	Air Conditioners	Water Purification Units
Tank & Pump Units	Mobile Field Kitchens	Refrig/Cold Storage Units
Bridge Boats	Laundry & Bath Units	Herman Nelson Heaters

Highlights of other Experience: Provided my industrial maintenance technology expertise while excelling in positions as a **POWER GENERATOR MECHANIC, HEAVY EQUIPMENT MECHANIC,** and **POWER PLANT MAINTENANCE MECHANIC** at locations throughout the Fort Bragg military installation.

PERSONAL

Excellent personal and professional references are available upon request.

Military products testing

Date

Exact Name of Person
Title or Position
Name of Company
Address
City, State, Zip

QUALITY CONTROL INSPECTOR

Dear Exact Name of Person: (or Dear Sir or Madam if answering a blind ad.)

Can you use an enthusiastic hard worker who offers experience in areas including warehousing and inventory control, supervising and counseling employees, and quality control?

I offer a reputation as a professional who excels in motivating and training others through "leadership by example" and a positive attitude. I have served my country in the U.S. Army while becoming known as a skilled technical inspector, inventory manager, and supervisor.

In my most recent job I earned two achievement medals for my performance in special project management and supervised seven people while handling inventory management activities and quality control actions. An earlier job allowed me opportunities in product marketing and testing as well as in working with civilian contractors to guarantee new products met government standards.

I hope you will welcome my call soon to arrange a brief meeting at your convenience to discuss your current and future needs and how I might serve them. Thank you in advance for your time.

Sincerely yours,

Howard Weston

Alternate last paragraph:

I hope you will call or write soon to suggest a time convenient for us to meet and discuss your current and future needs and how I might serve them. Thank you in advance for your time.

HOWARD WESTON

1110½ Hay Street, Fayetteville, NC 28305 • preppub@aol.com • (910) 483-6611

OBJECTIVE

To contribute to an organization that can use a hard worker who offers experience in inventory control and warehouse management along with proven strengths in supervising, motivating, and providing guidance to employees.

EXPERIENCE

Provide expertise as a technical instructor, supervisor, and inventory controller, U.S. Army, Fort Hood, TX:

QUALITY CONTROL INSPECTOR and INSTRUCTOR. (2005-present). Oversee seven employees, make the final determination of whether work was completed according to very strict guidelines, and control all aspects of inventory control from ordering, to receipt and storage, to issuance.

- Inspected packing of parachutes and equipment and made decisions directly affecting the lives and safety of thousands of paratroopers.
- Chosen to provide technical guidance to special projects in Mojave Desert, and Camp Pendleton, CA, trained combat divers who earned praise for their successes during the War on Terror.
- Exceeded requirements for stock levels of already "rigged," inspected, and approved parachutes.
- Created new record-keeping procedures which reduced paperwork.
- Honored with a commendation medal for "meritorious achievements," excelled as a specialist in managing aerial deliveries during a training project.
- Was handpicked to train more than 260 people in defenses against chemical warfare: several students used this information successfully while in Afghanistan and Iraq.

TEST AND MARKETING PROJECTS MANAGER. (2001-05). Refined my supervisory skills overseeing seven employees involved in preparing equipment and personnel for transport by air and in testing new products for possible military use.

- Participated in marketing and testing of products adopted for worldwide use, including the high-mobility multipurpose wheeled vehicle (HUMMWV).
- Contributed ideas and language used in producing technical manuals on the subjects of parachute rigging and preparing for air transportation. Packed 50 experimental systems and cargo packages used in NATO training and received an achievement medal.

Other experience:

WATER COMPANY TECHNICIAN. PA Water Company, Slippery Rock, PA (2000). Applied my mechanical skills by handling all of the town's water meter installations and also installed safety valves to aid in preventing excessive water use.

LAWN AND GARDEN MAINTENANCE SPECIALIST. Slippery Rock, PA (summers 1997-99). Learned to do preventive maintenance and repairs on lawn mowers, test swimming pool chemical levels, and properly maintain pools and lawns.

TRAINING

Excelled in approximately nine months of specialized training in these areas: leadership and supervisory skills; technical procedures for rigging parachutes

SPECIAL SKILLS

Offer skills in operating forklifts, power tools, weapons, commercial sewing machines, and vehicles up to five tons. Maintain weapons, commercial sewing machines, and wheeled vehicles.

PERSONAL

Was entrusted with a Secret security clearance. Am very skilled in personnel management, conducting training, and ordering and warehousing supplies.

Power production equipment

BRENT MITCHELL
1110½ Hay Street, Fayetteville, NC 28305 • preppub@aol.com
(910) 483-6611

QUALITY CONTROL INSPECTOR

OBJECTIVE

To contribute my outstanding technical power generation skills, knowledge of electrical and diesel equipment, and experience in testing and inspection.

AREAS of EXPERTISE & CLEARANCE

- Troubleshoot solid state devices including electromechanical devices.
- Experienced in diesel mechanics; read/interpret diagrams and schematics.
- Use equipment including phase rotation meters, energy analyzers, clamp-on amp meters, voltage continuity testers, portable load banks, maintenance tools, ohmmeters, voltmeters, analog and digital circuit testers, multimeters, and voltage probes.
- Entrusted with a **Secret** security clearance with SBI.

EDUCATION & TRAINING

Excelled in college-level training in 11 1/2 years of U.S. Air Force experience:

electrical power production
intrusion alarm detection
depot-level diesel generator maintenance
generator troubleshooting
effective training techniques
aircraft barrier systems
uninterruptible power systems
(ICS) diesel maintenance

EXPERIENCE

QUALITY CONTROL INSPECTOR. ACE Services International Company, Cairo, Egypt (2005-present). Perform preventive maintenance inspections in addition to operating and maintaining all prime, standby, and associated power plant equipment including four 300-KW White-Superior diesel generators supplying power to a remote radio station.

- Assist with three depot-level 10,000-hour as well as 1,000- and 4,000-hour inspections.
- Traced a problem that had caused four power outages within a month to its source, resolving the problem within 36 hours with no additional outages to date.
- Worked with a team discovering equipment defects in time to avoid costly engine overhauls and extensive downtime.
- Diagnosed a flaw in a regulator, saving both the equipment and crew from dangerous fluctuations in voltage levels.

SPECIAL EQUIPMENT TECHNICIAN. U.S. Air Force, Keflavik NAS, Iceland (2004-05). Inspected and maintained 16 intrusion detection systems providing security for sophisticated missile computer systems at a NATO facility.

- Installed and maintained low-voltage devices; tested/repaired/

calibrated relays, circuit breakers, sensors, and receivers; inspected for damage.

- Eliminated "nuisance" alarms by modifying a data transmission system.
- Maintained 98 MEP generator sets; responsible for three stand-by communication site generators.

POWER PRODUCTION SPECIALIST. U.S. Air Force, Keflavik NAS, Iceland (2000-04). Served as the Air Force's liaison with the Iceland Air Force on aircraft arresting system operation and maintenance; supervised five people while training personnel to use power generation systems.

- Prepared 38 generators for transport to Iraq during the War on Terror: completed the projected 72-hour job in only 48 hours.
- Implemented a highly effective maintenance scheduling system.
- Saved downtime and repair costs by rebuilding a defective auto switch relay and rewiring an engine auxiliary system.
- Played a major role in "excellent" ratings during a major inspection.

POWER PRODUCTION TECHNICIAN. U.S. Air Force, Kadena AB, Japan (1999-00). Maintained mobile diesel and gasoline-driven generator sets, conducted training, and was accountable for more than $400,000 worth of equipment.

- Repaired a critical aircraft barrier component, saving lives and aircraft.

POWER PLANT OPERATOR. U.S. Air Force, Kadena AB, Japan (1996-99). Operated/maintained a 1500-KW standby electric power plant and mobile diesel electric generator and kept all logs and records complete and up to date.

- Overhauled a 1,000-HP engine and a 750-KW Schoemaker generator.
- Was promoted to Power Plant Operator after excelling in maintaining and operating 5,000-KW Fairbanks and 800-KW Nuremburg generators; started and stopped engines; observed and interpreted instruments.
- Became skilled in troubleshooting while maintaining, repairing, and operating nine NTU 750-KW generators with switch gear and other equipment.
- During numerous outages, restored power in minimum time.

POWER PLANT GENERATION TECHNICIAN. U.S. Air Force, Heidelberg AIN, Germany (1994-96). Maintained and operated five 175-KW White-Superior diesel generators.

- Established new preventive maintenance guidelines: these standards led to a 99.9997% reliability rate for providing continuous power.

PERSONAL

Highly skilled troubleshooter. Will relocate worldwide. Work well with others.

Airframe and power plant repair

Date

Exact Name of Person
Title or Position
Name of Company
Address
City, State, Zip

QUALITY CONTROL SHOP SUPERVISOR

Dear Exact Name of Person: (or Dear Sir or Madam if answering a blind ad.)

I look forward to the opportunity to discuss with you soon how I could contribute to your organization through my reputation as a mechanically adept professional with outstanding leadership and motivational abilities.

While serving my country in the U.S. Army, I have advanced ahead of my peers and in less than five years have been handpicked for a job usually reserved for those with at least ten years of experience! In a letter recommending me for Warrant Officer Candidate School, I was described as "always the best among his peers." As a Quality Control Shop Supervisor, I am in charge of six Technical Inspectors while performing technical and shop inspections and monitoring maintenance programs for an organization with 24 OH-58D(I) Kiowa Warrior helicopters. In earlier jobs as a Helicopter Repair Technician, I was cited as one of the very best repairmen in my field. Additional experience has been in functional areas such as training personnel, maintaining records, weight and balance authority, accounting for multimillion-dollar assets, and overseeing utilization of communications equipment.

Highly motivated and always seeking ways to advance my knowledge and improve my skills, I have excelled in numerous rigorous training courses and have received respected certifications. Certifications I hold include the FAA Airframe and Power Plant Mechanic's rating; the ASE (Automotive Service Excellence) certification in engine repair, brakes and suspension/steering; and State of TX recognition as a Certified Inspector by the Commissioner of Motor Vehicles.

I am confident that I offer mechanical abilities, leadership skills, and the drive and determination to excel that will allow me to make major contributions to an organization in need of a talented and well-rounded professional. I can provide outstanding personal and professional references.

I hope you will welcome my call soon to arrange a brief meeting to discuss your current and future needs and how I might serve them. Thank you in advance for your time.

Sincerely,

Kenneth Watson

KENNETH WATSON

1110½ Hay Street, Fayetteville, NC 28305 • preppub@aol.com • (910) 483-6611

OBJECTIVE

To offer a reputation as a mechanically adept professional known for the ability to quickly and easily master new procedures, concepts, and equipment while excelling in providing leadership and setting the standard through my example.

EXPERIENCE

Advanced ahead of my peers to hold highly critical and visible technical and leadership roles while serving my country in the U.S. Army:

QUALITY CONTROL SHOP SUPERVISOR. Fort Bliss, TX (2005-present). Handpicked for the role of Quality Control (QC) Shop Supervisor, performed technical and shop inspections as well as monitoring QC maintenance programs for 24 OH-58D(I) Kiowa Warrior helicopters.

- In charge of six Technical Inspectors, have reached a level of rank and management position within five years that, on average, requires ten years of experience!
- Reorganized the QC Shop to include historical records, weights and balances, and technical files; earned commendable ratings for this effort during a critical inspection.
- Initiated a Corrosion Control Program for 24 aircraft, and conducted technical training programs related to maintenance, supply, and safety techniques.
- Maintained operational readiness rates above 80% for one fiscal year.
- Was cited in official performance evaluations as an especially skilled trainer who is always looking for new ways to motivate and train my subordinates.

HELICOPTER REPAIR TECHNICIAN. Fort Bliss, TX (2001-05). Supervised and evaluated the performance of eight subordinates while overseeing maintenance activities and service scheduling on eight helicopters valued in excess of $32 million; gained expertise in maintaining the following helicopters:

OH58D(I) Kiowa Warrior	OH-58C
UH-60E Blackhawk	AH-64C Apache

- Cited by technical inspectors and test pilots as one of the best repairmen they had ever worked with, was commended on one occasion for replacing an engine (normally a three-person job) in 24 hours. Maintained and accounted for in excess of $32 million worth of assets including helicopters and related equipment.
- Specially selected as Communications NCO responsible for the initial set-up of the aviation company's Tactical Operations Center, became highly proficient in the use of UHF, VHF, and SINCGAR radios; maintained perfect accountability of $155,000 worth of secure communications equipment.

EDUCATION & TRAINING

A.A. degree in General Studies from Central Michigan University, 2005.
Completed numerous executive development courses in management and supervision.
Excelled in the following technical training programs sponsored by the U.S. Army:
BNCOC: Learned skills of advanced leadership and perfected my technical inspection procedures. **Army Combat Lifesaver Course**: Completed training in first aid and medical procedures including initiating an intravenous infusion; this knowledge permitted me to save a soldier's life who had lost consciousness due to dehydration. **OH-58D Helicopter Repairer Course**: Assisted instructors in teaching maintenance for Kiowa Warrior aircraft.

RATINGS

FAA Airframe and Power Plant mechanic's rating, License #12345678.
ASE (Automotive Service Excellence) certified in engine repair, suspension and steering, and brakes with certification valid until 2006.
Certified Inspector, Commissioner of Motor Vehicles, the State of Texas, 2005.

PERSONAL

Happily married with two children. Non-smoker. Excellent physical condition. 27 years old.

Civil engineering projects

Date

Exact Name of Person
Title or Position
Name of Company
Address
City, State, Zip

QUALITY CONTROL SPECIALIST

Dear Exact Name of Person: (or Dear Sir or Madam if answering a blind ad.)

Can you use an enthusiastic and energetic young professional who offers a strong customer service orientation along with a work history which includes experience related to quality control and contract management?

While working for the Department of Transportation, I earned a reputation as a talented performer who could be counted on to go the extra mile and not quit until I found a way to get the job done. As a Production Control Specialist for civil engineering operations, I prioritized work requests, inspected work sites, and provided support by coordinating human and material requirements as well as performing data entry functions. I have consistently been described as being capable of handling details so that each job is coordinated properly and completed on time.

I am familiar with business software including Microsoft Word, PowerPoint, Excel, and Adobe PageMaker. I have a great deal of experience in preparing and presenting briefings and am accustomed to dealing with people ranging from city government officials, to civil engineers, to supervisory personnel, to a variety of mechanics.

A highly motivated individual with a reputation for honesty and integrity, I excel in finding ways to streamline daily operations and bring about improvements in productivity, customer service, and efficiency.

I hope you will welcome my call soon to arrange a brief meeting at your convenience to discuss your current and future needs and how I might serve them. Thank you in advance for your time.

Sincerely yours,

Laverna Moses

Alternate last paragraph:
I hope you will call or write me soon to suggest a time convenient for us to meet and discuss your current and future needs and how I might serve them. Thank you in advance for your time.

LAVERNA MOSES

1110½ Hay Street, Fayetteville, NC 28305 • preppub@aol.com • (910) 483-6611

OBJECTIVE

To contribute through a versatile background which includes outstanding performance in customer service, job scheduling and planning, and office operations as well as through my reputation as a dedicated, enthusiastic, and creative young professional.

EXPERIENCE

QUALITY CONTROL SPECIALIST. *Consistently earned praise for my ability to remain in control while overseeing the details of scheduling and coordinating multiple simultaneous civil engineering work requests and projects, Department of Transportation:*

Hill AFB, UT (2005-present). Prioritized work requests and entered essential data into computer records while ensuring that routine and emergency service calls were responded to promptly and that activities were tracked and recorded accurately.

- Initiated a series of computer reports which streamlined operations by allowing foremen to analyze potential problem areas and see that personnel were assigned in a way which maximized their particular skills.
- Assure that $775,000 worth of work was carried out by reorganizing the work order priority program and then maintaining the proper computer records.
- Prepare and organize slides and other materials used in weekly operations briefings.
- Set up and manage regular weekly work request review meetings.
- Was cited for my professionalism in controlling and allocating emergency human and material resources in response to a major aircraft accident.

Myrtle Beach, SC (2002-04). Received a commendation medal for my efforts in overseeing activities related to preparing a military complex for closure: worked closely with officials from the city while monitoring more than $1 million worth of contracts which studied the potential for civilian reuse of the facilities.

- Managed more than $500,000 in preliminary engineering studies.

Italy (2000-02). Was evaluated as being someone who always took the "extra steps beyond expected performance standards" which resulted in reducing problems and improving customer service in a civil engineering organization; maintained strict attention to detail in order to ensure work requests were completed in a timely and efficient manner.

- Contributed to the "excellent" rating received during a NATO evaluation by finding ways to eliminate conflicting information and control communication lines.

Pope AFB, NC (1998-00). Became skilled in taking care of the details of answering service calls, entering appropriate information into computer records, prioritizing requests, dispatching work crews, and keeping personnel informed of changing priorities and schedules.

- Controlled communications which allowed flood damage repairs at a recreational area to be coordinated and completed in the minimum amount of time. Used grid maps to plot damage control requirements and coordinate with fire, security, hazardous material, medical, explosive ordnance, and bioenvironmental personnel for emergency response.

EDUCATION & TRAINING

Completed one year of general studies, University of Maryland, campuses worldwide. Excelled in training programs including one leading to a certificate in Production Control, a Total Quality Management (TQM) course, and a leadership/management school.

- Earned a prestigious award as the top student in the leadership/management school and graduated in the top 2% from the Production Control class.

PERSONAL

Secret security clearance. Understand heating, air conditioning, plumbing, carpentry, electrical, and sheet metal work from eight years of exposure to civil engineering activities.

Appliances and electrical products

Date

Exact Name of Person
Title or Position
Name of Company
Address
City, State, Zip

QUALITY CONTROL SUPERVISOR

Dear Exact Name of Person: (or Dear Sir or Madam if answering a blind ad.)

I would appreciate an opportunity to talk with you soon about how I could contribute to your organization through my experience and knowledge in the areas of quality control operations management and production supervision.

As you will see by my resume, since 1998 I have been employed with Lavengood Industries, Inc., where I advanced from Production Supervisor to Quality Control Manager. During my three years in quality control, I established a series of classes which led to increased productivity and efficiency in the department by producing better-informed employees. I am knowledgeable of requirements for every stage of production, from fabrication to final assembly.

I offer strong mechanical abilities, training as a machinist and mechanical draftsman, and experience with CAD systems. I feel that these practical skills combine with my decision-making, leadership, and motivational abilities to make me a well-rounded professional with valuable assets sure to make me an important contributor to team efforts.

Although I am held in high regard by my current employer, I am in the process of relocating to your area and am selectively exploring employment opportunities with leading firms. I can provide outstanding references at the appropriate time.

I hope you will call or write soon to suggest a time convenient for us to meet and discuss your current and future needs and how I might serve them. Thank you in advance for your time.

Sincerely yours,

Brad Wilkins

BRAD WILKINS
1110½ Hay Street, Fayetteville, NC 28305 • preppub@aol.com • (910) 483-6611

OBJECTIVE

To apply my experience related to quality control and production supervision to an organization that can use my mechanical abilities as well as my proven skills in motivating employees, team building, and planning and scheduling work for maximum productivity.

EXPERIENCE

Advanced in supervisory and managerial roles within a manufacturing firm, Lavengood Industries, Inc., Jersey City, NJ:

QUALITY CONTROL SUPERVISOR. (2005-present). Apply my knowledge of quality control procedures to increase the department's efficiency and effectiveness for this manufacturer of ceiling fans, cooking range hoods, and other electrical items.

- Established communications with personnel at Hudson County Community College which led to the development of a successful series of classes for inspectors and other concerned company employees.
- Made improvements which led to significant decreases in the number of rejected parts, accomplishing this through improved training and communication between employees and supervisors.
- Receive products from vendors and examine their quality.
- Inspected parts at every stage from fabrication through final assembly.
- Work closely with engineering department personnel to coordinate for future changes and to correct items which were receiving complaints after being used.
- Excelled in communicating with personnel at all levels from production workers to management.

PRODUCTION SUPERVISOR. (1998-05). Was promoted to Quality Control Manager on the basis of my knowledge and accomplishments while making production schedules based on due dates and amounts of inventory on hand.

- Through increased communication and motivation, brought the department's efficiency rating to 98% from previous rates only in the 60-65% range.
- Established a system of regular weekly meetings which gave supervisors the opportunity to inform employees of requirements and production status.
- Was known for my ability to take workers who were not reaching their full potential and turning them into highly effective teams. Received "Supervisor of the Year Award" in 2003 for "outstanding dedication and loyal service and high ideals of professionalism."

Highlights of other experience:

- Learned responsibility and respect for authority as a Parachute Rigger supporting elements of the 10th Mountain Division at Fort Drum, NY. Gained supervisory experience while setting up machines and assisting machine operators at a manufacturing plant.

EDUCATION

Completed the **Mechanical Drafting** program and studied to be a **Machinist**, Hudson County Community College, Jersey City, NJ.
Attended training programs in areas including human relations and statistical quality control as well as seminars on "Staying Union Free."

TECHNICAL SKILLS

Can operate CAD (Computer-Aided Drafting) systems.
Am able to prepare clear, complete, and accurate working plans and detailed drawings, from rough or detailed sketches or notes, for engineering or manufacturing purposes.

PERSONAL

Offer strong organizational skills and have the ability to make well-informed decisions quickly and accurately. Am very effective in motivating others to work together as a team.

Avionics engineering

Date

Exact Name of Person
Title or Position
Name of Company
Address
City, State, Zip

QUALITY CONTROL SUPERVISOR

Dear Exact Name of Person: (or Dear Sir or Madam if answering a blind ad.)

I would appreciate an opportunity to talk with you soon about how I could contribute to your organization through my exceptionally strong problem-solving skills as well as my versatile management experience related to human resources, quality control, budgeting and finance, logistics and supply, transportation, and security.

Expertise related to production and supply

As you will see from my resume, while being promoted to the rank of Major in the U.S. Air Force, I have excelled in jobs which required unusual resourcefulness, creative strategic thinking, and prudent decision making. In my current job directing the highest-quality maintenance production and quality control ever achieved by my organization, I have continuously unplugged bottlenecks and solved stubborn problems in supply and logistics related to parts required to maintain and fly B-52 and KC-135 aircraft. With a reputation as an astute thinker and outstanding writer, I authored and published a 100-page "maintenance award package" named "best in the Air Force" and is now considered the model for others.

Strong background in quality control

While managing maintenance on a fleet of aircraft, I have learned to work in an environment in which there is "no room for error" so concepts like quality control and total quality management are truly "second nature" to me. Even in an earlier job as Chief of Administration for a 6,000-person Air Force base, I became known for my strong quality control orientation as I led that 41-year-old Air Force community to achieve an historical "first": it achieved top scores on three consecutive inspections of all areas of its operation.

Skills in human resources administration

I have also excelled in top human resources jobs. In one job I became the leader of a newly consolidated headquarters section supporting more than 600 people, and I rapidly implemented many novel ideas which boosted morale and productivity. On another occasion, I took over a marginal administrative operation and quickly turned it into a highly motivated operation.

You would find me to be a congenial and dynamic professional who is known for selfless dedication to duty. I can provide outstanding personal and professional references upon your request: I hope you will welcome my call soon when I try to arrange a brief meeting at your convenience to discuss your needs and how I might serve them. Thank you in advance for your time.

Sincerely yours,

Jenna Shortt

JENNA SHORTT

1110½ Hay Street, Fayetteville, NC 28305 • preppub@aol.com • (910) 483-6611

OBJECTIVE

To contribute to an organization that can use a dynamic self-starter known for vision and initiative who offers expertise related to quality control and total quality management, human resources and personnel supervision, as well as supply and maintenance administration.

CLEARANCE

Top Secret security clearance; familiar with security requirements pertinent to ensuring safe transportation and protection of aircraft, munitions, VIP's, and electronic assets.

EDUCATION

B.S. degree in Psychology/Sociology, University of Massachusetts, 1992.
Completed course work toward a Master's degree while excelling in full-time jobs as a military officer; also completed training related to human resources, quality control, finance/budgeting, total quality management (TQM), supply, and other areas.

EXPERIENCE

QUALITY CONTROL SUPERVISOR. U.S. Air Force, locations in the U.S. and Europe (2003-present). Managed up to 300 people while directing the highest-quality maintenance of B-52 and KC-135 aircraft this organization has ever had; planned and administered budgets ranging from $200,000 to $400,000 while being responsible for assets valued at millions of dollars; continually maximized sortie production without compromising safety or quality.

- On my own initiative, directed an effort to create a database of failure rates for avionics parts which we tracked for one year; this resulted in cost savings while improving parts availability through significant revisions made in parts ordered. Negotiated to obtain vehicles for munitions personnel and a fuel cell facility for repairing B-52G/H aircraft.
- Was evaluated in writing as "a very ingenious problem solver with superb self-initiative and a can-do attitude" and was praised for "a management style that promotes 'lessons learned' in order to prevent errors and promote improvements."
- As Officer in Charge of a Tanker Branch, in just five months unplugged stubborn bottlenecks, identified/corrected production problems, and implemented management innovations contributing to winning the 2004 "Best in SAC" aircraft maintenance award.
- In only 30 days, authored and published a 100-page "maintenance award package" that won competitive honors and is now considered the "model."

HUMAN RESOURCES ADMINISTRATOR. U.S. Air Force, Hill AFB, UT (2002). Was handpicked by the base commander to direct all administrative operations and personnel programs of this newly consolidated headquarters section—the first ever in SAC—supporting over 600 people in 17 diverse organizations.

- Established highly successful on-the-job training and self-inspection programs.
- Implemented many new morale-building ideas including a monthly recognition ceremony, a routine "gripe session," and an anonymous driver program with volunteers to pick up impaired individuals at any time of the day or night with no repercussions.

CHIEF OF ADMINISTRATION. U.S. Air Force, Hill AFB, UT (2000-02). Worked directly for the base commander of this Air Force community while overseeing all administrative functions for its 6,000 residents.

- Became known for my exceptional emphasis on quality control and "self inspection" as I led this 41-year-old base to receive the highest-possible ratings on three consecutive inspections of personnel administration, civil engineering, finance, and all other areas.

PERSONAL

Am known as an extremely mission-oriented and selfless individual who will find a way to achieve the goal. Am an analytical and logical thinker who has also been called a "visionary" because of my ability to anticipate future pitfalls and formulate practical strategic plans.

Educational services and programs

Date

Exact Name of Person
Title or Position
Name of Company
Address
City, State, Zip

QUALITY DIRECTOR, EXCEPTIONAL CHILDREN'S PROGRAMS

Dear Exact Name of Person: (or Dear Sir or Madam if answering a blind ad.)

With the enclosed materials, I am making formal application to the Ed.D. in Educational Leadership Program at University of Nevada.

You will see from my enclosed resume that I have an M.ED in Special Education and have gained a statewide reputation as an expert in the field of exceptional children's programs. In my current job with Webster County Schools, I supervise 95 teachers, teacher assistants, speech pathologists, and psychologists while directing the day-to-day operations of programs for special populations of children.

I was quite excited to see a question on the application related to my work in a program of significance with public schools, and I have written about my experience as a change agent implementing a new educational approach called Combined Education within the Webster County Schools.

I am committed to spending the rest of my life involved in the design and implementation of programs which will be "user-friendly" to teachers and students in the public schools, and I offer a "track record" of contributions to educational development within Nevada. I have been active in developing programs for transition from school to work for handicapped students, and I strongly believe the handicapped can be prepared and trained to become productive employees.

I would be delighted to make myself available to you for a personal interview, if you feel this is desirable or necessary. I am confident that I could become a distinguished alumnus of the doctoral program at UN, and I feel certain that my experience in teaching and administration would enrich the learning environment of the other doctoral students. Thank you in advance for giving my application every consideration.

Sincerely yours,

Nancy J. Vetstein

NANCY J. VETSTEIN

1110½ Hay Street, Fayetteville, NC 28305 • preppub@aol.com • (910) 483-6611

OBJECTIVE

To apply to the Doctoral Program (Ed.D.) in Educational Leadership in order to gain insight and knowledge that will enable me to continue, at even higher levels of leadership, my "track record" of contributions to educational development within Nevada and the nation.

EDUCATION

Certificate in Educational Administration and Supervision, University of Nevada, 1986.
M.ED. in Special Education, University of Nevada, Reno, NV, SC, 1974.
B.A. in History and Art, University of Nevada, 1971.
Extensive continuing education in areas related to Combined Education, Strategies for Teaching SLD Students, Mastery Learning, and other areas.

AFFILIATIONS & COMMUNITY LEADERSHIP

- Member, Liaison Committee, Kibler Mental Health Agency
- Active supporter, CARE Center, Muscular Dystrophy Association
- Advisory Board Member, NV Partnership Training System, University of Nevada
- Member for 20 years, national and state Council for Exceptional Children

EXPERIENCE

QUALITY DIRECTOR, EXCEPTIONAL CHILDREN'S PROGRAMS. Webster County Schools, Reno, NV (1994-present). Was promoted from Supervisor to Director, and now manage 95 teachers, teacher assistants, speech pathologists, and psychologists while directing the day-to-day operations of programs for handicapped children.

- Developed innovative new workshops and training opportunities for teachers on writing IEPs and transition education; gained considerable experience in writing federal grants.
- Worked closely with the Office of Civil Rights in collecting data related to transportation.

SUPERVISOR, EXCEPTIONAL CHILDREN'S PROGRAMS. Webster County Schools, Reno, NV (1988-94). Became knowledgeable about laws governing Exceptional Children's Programs while coordinating staff development workshops, chairing the Administrative Placement Committee, supervising Educational Diagnostic Centers, and overseeing services provided through the Homebound/Hospital and Vision Impaired Programs.

INSTRUCTIONAL SPECIALIST. Webster County Schools, Reno, NV (1982-88). Supervised the Behavioral Emotionally Handicapped Programs which included training teachers in writing and implementing behavior management plans.

- Authored the Behavior Management System currently in use in Webster County.

COLLEGE INSTRUCTOR. University of Nevada, Reno, NV (1979-80). On a part-time basis, taught the Introduction to Special Education Course, the Testing & Measurement Course, and the Gifted & Talented Course.

TEACHER, EXCEPTIONAL CHILDREN'S (EC) PROGRAMS. Webster County Schools Reno, NV (1971-79). At three different elementary schools, taught students with severe learning and behavioral problems; developed individual education programs for academic and behavioral instruction while completing diagnostic evaluations on each student.

- Saw first-hand that students ranked "at the bottom" can learn and grow with consistent expectations and genuine respect; later saw many of these students graduate from high school and become employed in the community.

PERSONAL

Outstanding references upon request. Proven ability to work in harmony with others.

Automotive and industrial parts

Date

Exact Name of Person
Title or Position
Name of Company
Address
City, State, Zip

QUALITY ENGINEER SUPERVISOR

Dear Exact Name of Person: (or Dear Sir or Madam if answering a blind ad.)

With the enclosed resume, I would like to express my interest in exploring employment opportunities with your organization.

As you will see from the enclosed resume, I am currently excelling as a Quality Engineer Supervisor, and I was promoted to my present position after working as a Quality Engineer. I am skilled at managing and implementing quality standards for manufacturing, packaging, and shipping of products worldwide, and I have initiated and implemented process improvements in multiple plants. In my current position, I assist eight manufacturing facilities in identifying and closing gaps utilizing the Business Excellence Process which focuses on customer satisfaction.

I began my career in industrial engineering after earning my B.S. in Mathematics, and as my career progressed, I decided to return to a university setting to earn my M.S. in Industrial Engineering while working full-time in a challenging management position.

Although I am held in the highest regard by my current employer and can provide outstanding references at the appropriate time, I am selectively exploring opportunities in other organizations. I am committed to the highest standards of quality assurance and customer satisfaction.

I hope you will call or write me soon to suggest a time convenient for us to meet and discuss your current and future needs and how I might serve them. Thank you in advance for your time.

Sincerely,

Victoria Mosley

Alternate last paragraph:
I hope you will welcome my call soon to arrange a brief meeting to discuss your current and future needs and how I might serve them. Thank you in advance for your time.

VICTORIA MOSLEY

1110½ Hay Street, Fayetteville, NC 28305 • preppub@aol.com • (910) 483-6611

OBJECTIVE

To contribute proven analytical, problem-solving, and communication abilities to an organization that can use a detail-oriented professional with excellent technical skills.

EDUCATION

Master of Science in **Industrial Engineering**, University of Detroit, MI, 2000.
- Inducted into Kappa Beta Alpha Industrial Engineering Honor Society.

Bachelor of Science in **Mathematics**, Marygrove College, Detroit, MI, 1995.

EXPERIENCE

Advanced with Great Lakes, Inc., Detroit, MI (2001-present):

2004-present: QUALITY ENGINEER SUPERVISOR. Promoted to manage and supervise three engineers, two technicians and 13 process auditors/receiving inspection employees while assisting plant personnel with quality issues related to manufacturing facilities, customer complaints, cost requirements, and delivery.
- Assist eight manufacturing facilities with closing gaps utilizing the Business Excellence Process which focuses on the customer and is similar to the M. Bridge Process.

2001-04: QUALITY ENGINEER IV, III, & II. Developed, managed, and maintained the quality system for Communication products; provided training to all personnel in ISO 9001 requirements and quality specifications while serving as Lead Assessor.
- Developed and implemented quality systems involving FMEAs, Control Plans, and Design Reviews to improve manufacturing processes for new products.
- Extensively involved in preventive and corrective actions taken during manufacturing processes; performed capability studies. Trained others in approving tooling changes and assisted with improvements in process engineering and error-proofing projects.

QUALITY CONTROL MANAGER. Detroit, MI (1999-01) and Ann Arbor, MI (1998-99). At one of the company's major plants, managed and implemented quality standards for manufacturing, packaging, and shipping of products in two hosiery facilities; a manufacturing facility with approximately 400 employees and a distribution center with more than 700 employees. Trained and supervised 23 quality control personnel.
- Worked closely with other manufacturing facilities and managers to correct problems with incoming products; helped operators and mechanics with process improvements.
- Initiated process improvements and error proofing activities in both plants.

Other experience:

QUALITY CONTROL SUPERVISOR. Great Lakes Telecommunications Cable Group, Flint, MI (1998). Managed the quality of single-mode optical fiber for telephone cables while attending graduate school; utilized the Team Concept to ensure fiber met quality standards set by Bell Laboratories.

OPERATIONS SUPERVISOR. Duracell Battery Company, Detroit, MI (1995-97). Led 50 personnel to meet demanding production and quality standards in the Raw Cell department of this giant corporation; controlled raw material inventory, scrap rates, and production costs using the Team Concept.

SPECIAL TRAINING

Have been certified by Great Lakes as a Lead Assessor for ISO 9001. Highly knowledgeable of the Demining Concepts, which figured heavily in my Masters thesis, as well as implementation of Total Quality Management and Team Building Concepts.

PERSONAL

Member of the American Society for Quality and Examiner for the Michigan M Bridge Process.

Automotive and industrial parts

Date

Exact Name of Person
Title or Position
Name of Company
Address
City, State, Zip

QUALITY INSPECTOR

Dear Exact Name of Person: (or Dear Sir or Madam if answering a blind ad.)

With the enclosed resume, I would like to make you aware of my extensive background in quality assurance and quality control.

As you will see from my resume, I became involved in quality management and quality inspection as a Communications Chief when I was responsible for ordering and controlling extensive inventories of equipment while also supervising 23 electronics specialists in calibrating, repairing, and troubleshooting communications-electronics equipment. In a subsequent job as Store Manager, I played a key role in helping a store earn a reputation for quality products by ensuring all equipment was in optimal condition and by performing quality inspections of all products purchased by the customers.

My philosophy that "Quality is our main concern," carried over to PRP Financial Services, where I managed stores in seven different states ensuring quality control techniques were implemented. I instilled in employees the concept that quality performance earns customer respect.

Most recently, in my dual roles at Goodyear Tire Company as a Side Wall Grinder and Inspector, I have taken great pride in knowing that no defective tires ever escaped my meticulous inspection.

I hope you will consider me for the position of Quality Control Manager which you recently advertised, and I would be honored if you would grant me an opportunity to interview in person for the position.

Yours sincerely,

Grant Shumaker

GRANT SHUMAKER

1110½ Hay Street, Fayetteville, NC 28305 • preppub@aol.com • (910) 483-6611

OBJECTIVE

To contribute to the Goodyear Tire Company through my extensive background related to quality inspection, quality assurance, and quality management.

EDUCATION

Received a Certificate from the two-year **Electrical Program**, The Colorado College, Colorado Springs, CO, 1992.
Completed more than two years of college-level training sponsored by the U.S. Army related to quality control, operations management, and electronic-communications, 1981-90.

EXPERIENCE

Since 1994, have excelled in these two jobs with the Goodyear Tire Company:
2005-present: QUALITY INSPECTOR. Take pride in the fact that I expertly inspect tires for serviceability to assure that no tires with defects escape my meticulous inspection.

- Am continuously aware that the customer is the last inspector, and I pride myself on looking at every tire through the customer's eyes; am a technical expert in all aspects of tire inspection.
- Work with the classifier and play a key role in correcting any recurring faults which I detect; am known for my excellent working relationship with all Goodyear personnel.

1994-05: SIDE WALL GRINDER. Learned first-hand that poorly finished tires will not sell, and also learned how poor work causes scrap which in turn costs the company money.

- Worked hard to instill in my fellow employees the desire to "do a job right the first time."
- Strongly believe that I was one of the best-ever Side Wall Grinders in the company; always took the view that I was trying to make a "lifetime repeat customer" out of the tire on which I was working on.

GENERAL MANAGER OF SALES. PRP Financial Services, Colorado Springs, CO (1992-94). Supervised 30 employees while managing stores in seven different states, and applied quality control techniques while working with finance companies all over the U.S.

- Instituted numerous new internal procedures which improved the quality of paperwork and which improved customer satisfaction because of better turnaround times.
- Instilled in employees the philosophy that "Quality is our main concern," and saw first-hand that a commitment to quality performance by employees earns customer respect.

STORE MANAGER. Colorado Springs Electronics, Colorado Springs, CO (1990-92). Played a key role helping this store earn a reputation for quality products; carried a product line which included stereos, home entertainment equipment, computers, televisions, records, and other equipment. Performed quality inspections of all products before customers took them out of the store, and discovered that quality emphasis was the key to lasting relationships.

GENERAL MANAGER. U.S. Army, Fort Carson, CO (1989-90). Was promoted ahead of my peers because of my excellent communication and supervisory skills; was a First Sergeant in charge of a 217-person unit. Molded inexperienced young soldiers into men and women committed to quality performance standards.

COMMUNICATIONS CHIEF. U.S. Army, Fort Carson, CO (1981-89). Planned and administered budgets for repair parts and tools, ordered and controlled extensive inventories of equipment, and supervised 23 electronics specialists.

PERSONAL

Have learned how to implement quality standards in service organizations as well as in companies that make a product. Offer an attitude of meticulous attention to detail.

Biomedical equipment

Date

Exact Name of Person
Title or Position
Name of Company
Address
City, State, Zip

QUALITY MAINTENANCE SUPERVISOR

Dear Exact Name of Person: (or Dear Sir or Madam if answering a blind ad.)

I would appreciate an opportunity to talk with you soon about how I could contribute to your organization through my experience and skills in the specialized field of biomedical equipment maintenance and repair as well as through my outstanding planning, organizational, and supervisory skills.

As you will see from my enclosed resume, I am a Medical Equipment Maintenance Services Supervisor and have been recognized with several medals and awards for my accomplishments and professionalism. I received extensive training which included the U.S. Army's year-long Biomedical Equipment Repair Course as well as additional training in information systems and PC maintenance and repair.

Presently overseeing maintenance for an inventory of more than 500 pieces of equipment, I was selected for this position and have been credited with exceeding expectations in all areas of operations. I have transformed an inefficient operation into one which is known for being customer friendly, productive, and efficient. I took the initiative to locate sources of free parts and equipment and obtained more than $200,000 worth of equipment at no cost to the organization. I have achieved additional success in several overseas assignments with task forces in Bosnia and Haiti where I have often been the only biomedical equipment technician servicing multiple locations and developing outstanding services.

If you can use an experienced professional with technical and mechanical skills along with well-developed managerial and leadership abilities, I hope you will contact me to suggest a time when we might meet to discuss your needs. I can assure you in advance that I could rapidly become an asset to your organization.

Sincerely,

Ward McCally

WARD McCALLY

1110½ Hay Street, Fayetteville, NC 28305 • preppub@aol.com • (910) 483-6611

OBJECTIVE

To contribute to an organization that can benefit from my experience and knowledge in the specialized area of biomedical equipment technology along with my excellent technical troubleshooting skills and my ability to motivate and lead others to achieve results.

EDUCATION & TRAINING

Excelled in extensive training which has included the Army's year-long Biomedical Equipment Repair Course as well as the following programs and courses:

- automated maintenance management information systems for equipment
- radiation protection
- digital film system/ teleradiology and video teleconferencing driver training

TECHNICAL SKILLS

Troubleshoot to the component level, medical equipment including (but not limited to) defibrillators, anesthesia, X-ray, ventilators, monitors, suction apparatus, and lab equipment.

EXPERIENCE

Have earned a reputation for my initiative and dedication to excellence as well as for my leadership, technical, and mechanical skills while serving in the U.S. Army:

QUALITY MAINTENANCE SUPERVISOR. Ft. Benning, GA (2000-present). Recognized with several medals and awards for my expertise and versatility; independently handle the details of developing and overseeing all aspects of maintenance support for more than 500 pieces of equipment; supervise one technician.

- Was credited with transforming an inefficient and unproductive section into one which is recognized as being customer service oriented and highly productive.
- On my own initiative, located sources and obtained more than $200,000 worth of parts and test equipment at no cost to the organization.
- Prepare annual evaluation reports for one junior manager as well as monthly counseling statements for all employees under my command.
- Schedule and perform preventive maintenance, electrical safety tests, and calibration.
- Oversee and personally maintain automated records of equipment inventories.
- Control over 300 types of repair parts to include inventorying, ordering, and stocking.
- Earned Army Commendation and Achievement Medals for my accomplishments supporting the quality of health care for personnel providing humanitarian assistance in Haiti: trained local personnel in electrical wiring and in equipment maintenance and helped establish a maintenance program in a new dental clinic.
- Received my fourth Army Commendation Medal and a NATO Medal for serving in Bosnia as the lone medical maintenance support specialist for several facilities: participated in multinational support efforts and developed effective training and a calibration schedule which increased levels of support and service provided by a medial task force.

BIOMEDICAL EQUIPMENT MAINTENANCE SUPERVISOR. Ft. Bragg, NC (1997-00). Supervised four people while providing outstanding support maintenance for an 80,000-member organization with its worldwide fast-response mission.

- Recognized as a subject matter expert, was called on to train other technicians.
- Trained and set up the first forward contact team in preparation for a major project in Africa.
- Designed and built a shelving system for repair parts which significantly improved storage.

BIOMEDICAL EQUIPMENT TECHNICIAN. Ft. Bragg, NC (1995-96). Learned depot-level repair procedures and earned an Army Commendation Medal.

PERSONAL

Studied college-level English and History at the college level.

Call center management

Date

Exact Name of Person
Title or Position
Name of Company
Address
City, State, Zip

QUALITY SPECIALIST & PROJECT TEAM LEADER

Dear Exact Name of Person: (or Dear Sir or Madam if answering a blind ad.)

With the enclosed resume, I would like to make you aware of my interest in exploring employment opportunities with your organization and introduce you to my extensive customer service and quality assurance experience.

My customer service training began very early in my life, when I worked as a Customer Service Specialist in high school. As a high school senior, I worked up to 40 hours weekly on a 4pm-1am shift as I provided Visa cardholders with information about their Membership Awards Program. After relocating to Watertown, NY, with my military father and our family, I worked as a Client Relations Specialist for Dell in a newly opened store. Based on my outstanding performance, I was offered a job with Dell when I relocated back to my hometown of Columbia, SC.

Instead of continuing with Dell, however, I accepted a position with Omaha Mutual as a Bilingual Service Specialist, and I excelled in all aspects of my job while providing customer service support in English and Spanish. In 2003, Omaha Mutual promoted me to the position of Quality Specialist & Project Team Leader in a call center recognized by J.D. Powers and Associates for "call center operation customer satisfaction excellence." As one of only five Quality Specialists overseeing a 427-person call center, I play a key role in assuring the highest level of customer service for Omaha Mutual's 30 million customers. I have excelled in numerous professional development and executive training programs, and I have completed examinations in a variety of areas ranging from customer service to insurance company operations.

What I have discovered in my customer service experience is that the highest level of customer satisfaction is achieved through teamwork and through the creative application of outstanding communication and problem-solving skills. While solving problems and identifying opportunities, I utilize a variety of analytical tools and software including Excel spreadsheets, Access databases, written reports, graphs, and presentation software. I offer a proven ability to work well independently, as a team member, or in a leadership role. In addition to my strong and genuine commitment to customer satisfaction, I possess outstanding written and oral communication skills.

If you can use a cheerful and insightful individual who offers a proven ability to work well with others, I hope you will contact me to suggest a time when we might discuss your needs. I can provide outstanding personal and professional references at the appropriate time. Thank you in advance for your time and professional courtesies!

Yours sincerely,

Lorena Santiago

LORENA SANTIAGO

1110½ Hay Street, Fayetteville, NC 28305 • preppub@aol.com • (910) 483-6611

OBJECTIVE

To offer my customer service and quality assurance experience, along with strong computer skills, to a company that can use a skilled communicator and problem solver who offers the ability to profitably impact a company's bottom line through my ability to ensure high levels of customer satisfaction.

LANGUAGE

Fluently speak, read, and write **Spanish;** volunteered as a Translator for Special Olympics events and with the U.S. Army, Fort Jackson, SC.

COMPUTERS

Proficient with Microsoft XP (Word, Excel, PowerPoint, Visio, Outlook, Access) and Intranet.

EDUCATION

Completing **B.S. degree in Chemistry,** University of South Carolina, Columbia, SC. Am pursuing this degree in my spare time while excelling my demanding full-time position.

Previously completed **30 hours of course work pursuing A.A. degree,** Jefferson Community College, Watertown, NY. Attended college in the mornings and worked from noon to 9 p.m.

Graduated from Columbia High School, Columbia, SC, 1999.

- Selected by Junior ROTC instructors for a leadership role. Junior and Senior Class **President.** Honored as **Salutatorian** (2nd academically) of my graduating class; Honor Society member.

Completed computer, sales, and customer service training sponsored by Dell.

EXPERIENCE

Promoted to increasing responsibilities with Omaha Mutual Fund, Columbia, SC.

2003-present: QUALITY SPECIALIST & PROJECT TEAM LEADER. Was promoted to a critical quality position in a call center recognized by the J.D. Powers and Associates Certified Call Center Program (SM) for "customer satisfaction excellence."

- Have been commended for my strong analytical skills, superior verbal and written communication skills, and in-depth call monitoring experience while continually demonstrating my commitment to providing an outstanding customer service experience for Omaha Mutual's 30 million customers.
- In this 427-person call center, review procedures and processes and recommend enhancements to improve efficiency. Consult with management to develop and formulate long-term strategic plans to ensure Omaha Mutual's leadership in the area of providing the highest quality customer service.

2001-2003: BILINGUAL CUSTOMER SERVICE SPECIALIST. Excelled in all aspects of my job while providing customer service support in both English and Spanish.

CLIENT RELATIONS SPECIALIST. Dell, Inc., Watertown, NY (1999-2001). Earned recognition for my ability to apply technical training in a common sense approach while assisting customers in a newly opened store specializing in configuring computer systems to users' needs. Patiently yet aggressively resolved client issues when shipments were delayed due to configuration problems. Based on my excellent performance in this job, was offered a position with Dell in Columbia, SC, but chose to work for Omaha Mutual.

CUSTOMER SERVICE TRAINEE. American Services, Inc., Columbia, SC (1999). Excelled in an intensive customer service training program before relocating to Watertown, NY. Worked with financial institutions on client financing.

PERSONAL

Offer well-developed decision-making and problem-solving skills. Am a positive, energetic, and enthusiastic individual who can be counted on to get the job done. Outstanding references.

Motor control fabrication

Date

Exact Name of Person
Title or Position
Name of Company
Address
City, State, Zip

QUALITY TECHNICIAN

Dear Exact Name of Person: (or Dear Sir or Madam if answering a blind ad.)

With the enclosed resume, I would like to acquaint you with my skills as an experienced commercial and industrial electrician with a solid background in the wiring, maintenance, troubleshooting, and repair of electrical systems.

As you will see from my enclosed resume, I have extensive training in various types of wiring and electrical systems. In my most recent position at Gorham Industries, I assembled, wired, and performed troubleshooting on motor control units for industrial and commercial grade air compressors and chillers. Through our efforts, the department was awarded ISO 9002 certification while achieving the company's goal of "zero defects" in on-time shipping.

At Berenger Electric, I acquired valuable experience in new construction wiring and installation as well as applying my knowledge in reading blueprints and schematics while working on the new Morehouse Medical Center building.

During my years of experience, I have learned to troubleshoot and repair high-tension systems and electrical substations, as well as working with the wiring and electrical systems of a wide variety of ships and power equipment. In addition to the Electrical Wireman Journeyman's Course, I have completed supplementary training courses at Washington University and Payne Technical College. I feel that my extensive industry experience and education will be a strong asset to your organization.

If you can benefit from the services of a highly experienced electrician whose safety and quality assurance skills have been tested in a wide variety of industrial and commercial environments, please contact me to suggest a time when we might meet to discuss your needs and how I might meet them. Thank you in advance for your time and consideration.

Sincerely,

Roy L. Munster

ROY L. MUNSTER

1110½ Hay Street, Fayetteville, NC 28305 • preppub@aol.com • (910) 483-6611

OBJECTIVE

To benefit an organization that can use an experienced electrician with a solid background in wiring, maintenance, troubleshooting, and repair of industrial and commercial electrical systems.

EDUCATION

General Manufacturing (GMC) Course, Payne Technical College, 1998.
Zinger-Miller Course, Payne Technical College, 1998.
Electrical Equipment Testing & Maintenance, Washington University, 1989.
How to Charge Substation Batteries course, Canal Zone Vocational School, 1988.
Graduated from Canal Zone Vocational School's Electrical Wireman Journeyman's course, Panama.

SPECIAL SKILLS

Offer expertise in the following areas and with the following equipment:

troubleshooting all electrical devices
wire machines
heat treating equipment
plumbing
fabrication
3-phase 208 volt wye delta controls
Lincoln & Hobart welders
machine shop
electrical circuit design
AC/DC motors & generators
3-phase 480 volt wiring
electrical controls

EXPERIENCE

QUALITY TECHNICIAN. Gorham Industries, Payne, OR (2000-present). Perform various electrical assembly and wiring tasks in this fast-paced factory environment.

- Read wiring diagrams, schematics, and engineering order specifications to ensure units were completed according to requirements and proper standards.
- Assemble, wire, and perform troubleshooting on motor control units for industrial and commercial grade air compressors and chillers.
- Learned testing and troubleshooting of a wide variety of motor control units from different manufacturers.
- Was credited with making contributions which allowed the company to reach its goal of "zero defects" in on-time shipping.
- Member of a department which successfully qualified for ISO 9002 certification.

ELECTRICIAN. Berenger Electric Company, Payne, OR (1998-00). Primary responsibility was wiring and electrical installation of new construction at the Morehouse Medical Center job site.

- Installed junction boxes; mounted and wired electrical panels.
- Cut, bent, and installed electrical conduit of various sizes.
- Ran wires and cables through conduit to junction boxes, panels, and breaker boxes.

Other experience: WIREMAN/ELECTRICIAN. Panama Canal Zone, Balboa, Panama. Started with Panama Canal Zone as Electrician Apprentice at Canal Zone Vocational School; advanced to Wireman/Electrician after completing the program.

- Performed routine maintenance, troubleshooting, and repair on air and oil circuit breakers of electrical substations.
- Maintained various power equipment, including forklifts, high-capacity battery chargers, welding machines, magnetic drills, etc.
- Learned to troubleshoot and repair high-tension electrical systems.
- Worked on wiring, electrical systems, and batteries for ships, barges, beacons, and buoys.

PERSONAL

Excellent personal and professional references are available upon request.

Waste management

Date

Exact Name of Person
Title of Person
Name of Company
Address
City, State zip

RECYCLING GENERAL MANAGER

Dear Exact Name of Person: (Dear Sir or Madam: if answering a blind ad.)

I would appreciate an opportunity to talk with you soon about how I could contribute to your organization through my proven ability to develop cost-effective and popular solutions for the problems of commercial, residential, and industrial customers.

As you will see from my resume, I am currently managing a company in the solid waste and recycling industry. I have used my public relations and negotiating skills to gain valuable new commercial contracts, and the new incentive and production payroll programs I have established in various cities have benefitted the company as well as the most outstanding employees. In the past year I have nearly doubled the multimillion-dollar revenues of this corporation's largest division with no percentage increase in cost.

In a previous job I doubled revenues of that company and then organized and arranged the sale of the corporation to my current employer.

I believe strongly that cost-effective planning and problem solving are keys to satisfying and retaining customers, and I have become respected for my ability to develop innovative new solutions for "old" problems ranging from external recycling to internal employee productivity.

I hope you will welcome my call soon when I try to arrange a brief meeting at your convenience to discuss your current and future needs and how I might serve them. Thank you in advance for your time.

Yours sincerely,

Gilbert Thayers

Alternate last paragraph:

I hope you will call or write me soon to suggest a time when we might meet to discuss your current and future needs and how I might serve them. Thank you in advance for your time.

GILBERT THAYERS
1110½ Hay Street, Fayetteville, NC 28305 • preppub@aol.com • (910) 483-6611

OBJECTIVE

To contribute to an organization that can benefit from my strong planning, management, and public relations skills as well as from my proven ability to streamline operations, improve customer service, cut costs, and boost profit.

EDUCATION

B.A. degree in Business Management, University of Washington, Seattle, WA, 1994.

EXPERIENCE

GENERAL MANAGER. Northwestern Refuse Systems, Inc., Seattle, WA (2003-present). Began with NWRS, Inc. as **Operations Manager** in 2003; was promoted to **Special Project Coordinator** in 2004 and advanced to **General Manager** in 2005; since taking over as General Manager, have nearly doubled local company division revenue from its 2005 level of $3 million while establishing a new incentive payroll system and setting up a popular new recycling route for 10,000 customers at no additional cost to them.

- Lobbied for and negotiated a commercial contract with City of Seattle.
- Am in charge of this corporation's largest division operating 25 trucks serving 10,000 residential customers, 2,400 commercial customers, 400 industrial clients, and 400 portable toilet accounts; also manage seven clerical workers, 28 laborers/drivers, two sales representatives, and two supervisors.
- As **Special Project Coordinator** managed an NWRS operation in Spokane, WA, and handled extensive labor disputes there; established payroll incentive and production programs in Boise, ID; Pocatello, ID, Portland, OR; and Missoula, MT.
- As **Operations Manager**, supervised 28 field personnel and operated all equipment including Dempster, Heil, Lodal, Loadmaster, Pakmor, Galbreth rolloff, and marathon compactors; led this division to reduce needed trucks by 50% through effective rerouting; and reduced personnel from 40 to 28.
- Have become skilled in saving money through proper routing; have developed recycling routes in numerous cities.

MANAGER/SALES DIRECTOR. Washington Waste, Co., Seattle, WA (1996-03). Organized and arranged the sale of this company to NWRS Waste Management, LTD in July 2003 after doubling its revenue base; approved all sales contracts for this local company which served 8,000 residential and 375 commercial customers.

- Relied primarily on my planning, negotiating, and communication skills in dramatically boosting this company's profitability and net worth.
- Started "from scratch" a new commercial pickup and sales division.

MANAGER TRAINEE. PRP Services, Inc., Seattle, WA; Portland, OR; and Boise, ID (1994-96). After receiving extensive training related to landfill, office procedures, and customer service, assisted in the promotion of customer service agreements and gained experience in commercial route analysis.

- Excelled as a **Sales Representative;** opened a new sales area in Ada County, and headed up a study related to the conversion of lugger to roll off system.
- Acted as dispatcher, safety coordinator, and truck driving supervisor.
- Operated landfill heavy equipment and Rex compactors.

PERSONAL

Have a proven "knack" for developing and implementing cost-effective solutions to tough problems. Am skilled in managing personnel and customer relations.
Use Microsoft Word, Excel, and Access software.

Recycling industry

Date

Exact Name of Person
Title or Position
Name of Company
Address
City, State, Zip

RECYCLING PLANT MANAGER

Dear Exact Name of Person: (or Dear Sir or Madam if answering a blind ad.)

I would appreciate an opportunity to talk with you soon about how I could contribute to your organization through my solid waste management experience and my knowledge of inventory control, logistics, and distribution.

As you will see from my resume, I have had extensive experience in the recycling field. In my most recent position as a plant manager, I have sharpened my technical, administrative, and communication abilities.

My achievements as a plant manager have included opening a new plant which has advanced to become "number one" in productivity. This has been accomplished with a work force out of which 90% speak Spanish as a primary language.

I was "handpicked" out of six candidates for the plant manager position after only seven months as an assistant plant manager. In this job, my input and management were key factors in increasing output 20% while simultaneously reducing the number of employees needed per shift.

As a graduate of the U.S. Naval Academy, I also have earned my M.A. degree in Business and have excelled in specialized training for executives related to maintenance, supply, and logistics. While serving my country as a Marine Corps officer, I was selected for jobs normally reserved for more senior officers.

You would find me to be a bright and inquisitive professional who excels in training and motivating employees, both in groups and one-on-one. I hope you will welcome my call soon to arrange a brief meeting at your convenience to discuss your current and future needs and how I might serve them. Thank you in advance for your time.

Sincerely yours,

Christopher Maples

Alternate last paragraph:
I hope you will call or write me soon to suggest a time convenient for us to meet and discuss your current and future needs and how I might serve them. Thank you in advance for your time.

CHRISTOPHER MAPLES

1110½ Hay Street, Fayetteville, NC 28305 • preppub@aol.com • (910) 483-6611

OBJECTIVE

To contribute to an organization that can use an accomplished, resourceful industrial production professional who offers skills in solid waste management along with experience in recycling.

EDUCATION & TRAINING

M.A. in Business, Vanderbilt University, Nashville, TN, 2004.
B.S. in Physical Science, U.S. Naval Academy, Annapolis, MD, 1999.
Excelled in training for Marine Corps officers related to logistics, personnel management, and transportation management.

LICENSES

Licensed by the Virginia Department of Environmental Protection to supervise a material reclamation facility, valid until 2009.

EXPERIENCE

PLANT MANAGER. National Recycling Systems, Richmond, VA (2005-present). Earned promotion to plant manager after only seven months and supervise 12 operations and maintenance employees at this recycling company.

- Opened a new plant complying with DEP/OSHA regulations.
- Plan budgets to improve plant operations and profits.
- Maintain a "number one" ranking in productivity.
- Act as a liaison with community officials. Coordinate all hiring and firing activities.
- Troubleshoot and repair air classification systems, conveyors, and three phase motors.

ASSISTANT PLANT MANAGER. National Recycling Systems, Nashville, TN (2004). As "second in command" of this plant employing 60 people, supervised 24 production, maintenance, and quality control employees.

- Oversaw all production, quality control, shipping, and maintenance.
- Increased production 20% while eliminating an unnecessary sixth shift.
- Cross-trained employees, reducing each shift from 24 to 21 workers.
- Mastered the business and technical aspects of the recycling industry.

LOGISTICS CHIEF. U.S. Marine Corps, Camp Lejeune, NC (2003-04). Trained, motivated, and managed up to 17 people while directing inventory control and logistics support provided to a 1,200-person organization.

- Advised a chief executive on inventory, maintenance trends, and budgeting. Oversaw control of a diversified inventory and supervised the receiving and issuing functions.
- On my own initiative, drastically revised the organization's logistics policies and procedures: efficiency and customer morale soared. Streamlined/reorganized internal operations with the result that we improved quality and quantity of work performed with a leaner staff.

TRAINING DIRECTOR. U.S. Marine Corps, Camp Lejeune, NC (2003). Set new records for managerial performance in this job supervising the training of 175 people and evaluating both their performance and potential.

TRANSPORTATION CHIEF. U.S. Marine Corps, Camp Lejeune, NC (2001-02). Supervised the provision of vital transportation services for this 6,000- person community while training and managing up to 45 employees. Prepared a budget of up to $1.9 million.

- Created a recycling program that produced $1 million in annual revenue.

PERSONAL

Am a highly adaptable professional who enjoys solving problems.

Strategic and contingency planning

Date

Exact Name of Person
Title or Position
Name of Company
Address
City, State, Zip

RISK MANAGEMENT DIRECTOR

Dear Exact Name of Person: (or Dear Sir or Madam if answering a blind ad.)

With the enclosed resume, I would like to make you aware of my interest in exploring executive positions within your organization which could make use of my aviation expertise as well as my strong problem-solving, organizational, and management skills.

As you will see from my resume, I excelled in a track record of promotion within the U.S. Air Force while rising to the rank of Lieutenant Colonel. In my earliest jobs as a pilot, I was recognized for my initiative, judgment, knowledge, and leadership skills and was recommended for rapid advancement into supervisory and then into executive positions.

In one job, I was Chief of Branch Operations in charge of planning and directing strategic and operational airlift programs worldwide while managing airlift missions valued at $35 million. As an Operations Officer in Italy, I was praised as an "inspirational leader" while leading more than 430 aviators, maintenance personnel, and others involved in airlift missions supporting NATO and the Joint Chiefs of Staff. In a subsequent job as Chief of Safety at one of the nation's busiest airlift hubs, I directed mishap investigations while conceiving, developing, and implementing risk management, safety, and mishap prevention programs affecting thousands of people and multimillion-dollar aircraft.

With a reputation as a principled individual known for integrity, I always received the highest formal evaluations of my performance, and I was consistently described as a "model leader" with an exemplary work ethic and an enthusiastic management style. My problem-solving and decision-making skills have been refined while operating in environments in which there was "no room for error," and I have held one of our nation's highest security clearances (Top Secret SCI). I can provide outstanding personal and professional references.

If you can use a seasoned executive who has an outstanding reputation in the aviation industry, I hope you will contact me to suggest a time when we might meet to discuss your needs. I am single and can relocate worldwide according to your needs.

Sincerely,

Trevor Smith

TREVOR SMITH

1110½ Hay Street, Fayetteville, NC 28305 • preppub@aol.com • (910) 483-6611

OBJECTIVE

To contribute to the pursuits of an organization needing a polished communicator and resourceful, strategic thinker offering vision, leadership, and team-building skills developed during a distinguished career as a military officer in international environments.

EDUCATION

Master of Business Administration, University of Kentucky, Lexington, KY, 1988.
Bachelor of Science in Civil Engineering, Bellarmine College, Louisville, KY, 1981.

EXPERIENCE

Rose to the rank of Lieutenant Colonel while serving in the U.S. Air Force:
2005-present: RISK MANAGEMENT DIRECTOR. Cannon AFB, NM. Assist in managing a $36.8 million budget and coordinating the activities of more than 1,000 people in five diverse organizations; ensure that all personnel are highly trained and integrated to accomplish wing mission. Instructor pilot for 240 officer aviators.

- Institutionalized risk management throughout the operations group; principles utilized in every aspect of daily operations significantly improved safety.
- Led a runway extension project that will nearly double airlift capability; gained expertise in the Environmental Impact/Assessment Process and integrated civilian, environmental and military groups in "selling" the concept to the affected civilian communities.

2003-04: CHIEF OF SAFETY. Cannon AFB, NM. Conceived, developed, and implemented risk management, safety, and mishap prevention programs affecting 5,000 people, a fleet of 91 aircraft, and aircraft operations. Directly supervised 12 executives while overseeing safety programs in 32 diverse organizations. Instructor pilot.

- Ran a program rated "outstanding" by Headquarters Air Mobility Command: 26,000 aircraft sorties handling 120,000 passengers and 3,600 cargo tons, and 65 major construction projects with no accidents.
- Oversaw research of bird/aircraft strike data that resulted in elimination of costly, ineffective risk control measures for Air Mobility Command C-130 aircraft.
- Proposed risk management strategies for flight and maintenance operations were incorporated into major command operating policy.
- Developed a working knowledge of OSHA requirements and instituted Risk Management and safety programs throughout a large, diverse and fast-paced organization.

2002-03: OPERATIONS OFFICER. Aviano AB, Italy. Led squadron operations and maintenance of 19 combat-ready C-130 aircraft and over 430 personnel, including 85 officers implementing airlift operations throughout Europe, and the Middle East.

1999-01: BRANCH CHIEF. Washington, DC. Planned, directed and provided oversight as well as preparing, coordinating, and implementing U.S. Air Force policy for worldwide airlift operations in support of sensitive Department of Defense, U.S. government, and National Security Council programs. Managed $35 million airlift system budget servicing 75 countries.

1995-99: DIVISION CHIEF. Cannon AFB, NM. Directed one of two offices in the U.S. Air Force providing support and expertise on the Adverse Weather Aerial Delivery system (AWADS) to all commands. Oversaw training and standardization/evaluation for 400 crew members and a 26,000 flying hour program involving airlift, aeromedical evacuation, and maintenance activities worldwide.

PERSONAL

Completed training related to specialized areas including international terrorism, foreign internal defense, total quality management, gender and sensitivity, and training as a pilot.

Aviation maintenance

Date

Exact Name of Person
Title or Position
Name of Company
Address
City, State, Zip

SAFETY INSPECTOR

Dear Exact Name of Person: (or Dear Sir or Madam if answering a blind ad.)

With the enclosed resume, I would like to make you aware of my background as a versatile and experienced professional with a history of success in areas which include computer operations, logistics and supply, production control, and employee training and supervision.

As you will see, I am completing requirements for my bachelor's degree in Computer Science which I expect to receive this winter. I am proud of my accomplishment in completing this course of study while simultaneously meeting the demands of a career in the U.S. Army. In my current military assignment as a Technical Inspector at Fort Campbell, KY, I ensure the airworthiness of aircraft utilized by the 101st Airborne Division which is required to relocate anywhere in the world on extremely short notice in response to crisis situations.

Throughout my years of military service I have been singled out for jobs which have required the ability to quickly make sound decisions and have continually maximized resources while exceeding expected standards and performance guidelines. I have been responsible for certifying multimillion-dollar aircraft for flight service, training and supervising employees who have been highly productive and successful in their own careers, and applying technical computer knowledge in innovative ways which have further increased efficiency and productivity.

I am confident that I offer a combination of technical, managerial, and supervisory skills and a level of knowledge which will allow me to quickly achieve outstanding results in anything I attempt. Known for my energy and enthusiasm, I am a creative and talented professional with a strong desire to make a difference in whatever setting and environment I find myself.

I hope you will contact me to suggest a time when we might meet to discuss your needs. I can assure you in advance that I could rapidly become an asset to your organization.

Sincerely,

Gerald Fueller

GERALD FUELLER

1110½ Hay Street, Fayetteville, NC 28305 • preppub@aol.com • (910) 483-6611

OBJECTIVE

To offer a versatile background emphasizing computer operations, logistics, inventory control, production control, and security to an organization that can benefit from my experience as a supervisor and manager with a reputation for creativity.

EDUCATION & TRAINING

Will receive a **B.S. in Computer Science**, Hopkinsville Community College, KY, Fall 2005. Selected for extensive training, including courses for security managers and computer operations and applications as well as technical courses in aircraft repair and maintenance, professional leadership development, equal opportunity practices, and Airborne School.

COMPUTER SKILLS & CLEARANCE

Am highly computer literate and offer skills related to the following:

Word Excel PowerPoint Access

Was entrusted with a Top Secret security clearance.

EXPERIENCE

Earned a reputation as a focused and goal-oriented professional, U.S. Army:

SAFETY INSPECTOR. Fort Campbell, KY (2004-present). Officially cited for my in-depth knowledge of aviation maintenance and logistics issues, conduct technical inspections on 24 helicopters while also providing training, guidance, and supervision.

- Transformed a substandard quality control section into one which sets the example.
- Received the NATO Medal for Service for my contributions during operations in the Middle East in 2005.

SUPERVISOR FOR AVIATION MAINTENANCE AND SUPPLY. Fort Campbell, KY (2002-03). Made numerous important contributions while coordinating supply, maintenance, and readiness issues for units throughout the 101st Airborne Division.

- Received a letter of commendation from a three-star general for my expertise demonstrated as the Security Manager for the corps headquarters and for personally revitalizing the physical security program which received "commendable" ratings.
- Developed a database for managing aircraft maintenance histories which greatly increased the reliability of information about the organization's 968 assigned aircraft.
- Created and presented effective training on physical security and ADP operations.

MAINTENANCE SUPERVISOR. Fort Campbell, KY (2001-02). Provided oversight for a $6 million annual operating budget and ensured the quality and timeliness of all phases of support for a fleet of 104 aircraft while also reviewing daily/monthly status reports, coordinating supply up to the wholesale level, and recommending procedural changes.

- Consistently exceeded Department of the Army standards for aircraft availability and was singled out for praise for my expertise as a trainer, mentor, and leader.

MAINTENANCE SUPERVISOR. Italy (1998-01). Cited as directly responsible for a high level of achievement and productivity; collected and processed maintenance data on 123 aircraft while controlling flight safety information and support for four divisions.

- Selected to oversee a project during which new equipment and automation assets were integrated into use; achieved a smooth transition.
- Was a member of the team which first fielded the ULASS computer system which is a tool for managing production, maintenance hours, and supplies.

PERSONAL

Am known for my enthusiastic style of leadership and reputation for unwavering moral and ethical standards. Was honored with numerous Meritorious Service, Commendation, and Achievement Medals in recognition of my contributions and professionalism.

Construction and roofing

Date

Exact Name of Person
Title or Position
Name of Company
Address
City, State, Zip

SAFETY INSPECTOR & QUALITY CONTROL SUPERVISOR

Dear Exact Name of Person: (or Dear Sir or Madam if answering a blind ad.)

With the enclosed resume, I would like to make you aware of my interest in joining your company in some capacity which can utilize my extensive experience in safety, technical inspection, and quality control.

As you will see from my resume, I became skilled in the safety and quality control while working in the U.S. Army. I was promoted ahead of my peers from Crew Chief and Crew Leader to Technical Inspector, and I was entrusted with responsibility of inspecting helicopters, hangars, maintenance inventories, and other areas to assure compliance with safety and quality control guidelines and regulations.

After serving my country, I transferred my management and technical skills into the construction industry, and I have excelled in jobs as a Superintendent, Safety Inspector, and Quality Control Supervisor. I have completed U.S. Army Corps of Engineers training related to Construction Quality Management.

I am the recipient of numerous honors and awards for outstanding performance, and I can provide outstanding personal and professional references. I hope I will have the opportunity to meet with you in person to show you that I am a knowledgeable and hard-working individual who could become a valuable asset to your company.

Sincerely,

Alexander Thomas

ALEXANDER THOMAS

1110½ Hay Street, Fayetteville, NC 28305 • preppub@aol.com • (910) 483-6611

OBJECTIVE

Management position which can utilize my background in safety and quality control as well as my experience as a construction superintendent and technical inspector.

EDUCATION

Jefferson Community College: completed two years of studies in Horticulture, one year in Carpentry, and one year in Cabinetmaking.
U.S. Army: Completed extensive management and technical training including advanced courses for NCOs, MAP TOE, Safety Courses, and Technical Aircraft Inspection Courses.
U.S. Army Corps of Engineers: Completed Construction Quality Management training.

EXPERIENCE

SAFETY INSPECTOR & QUALITY CONTROL SUPERVISOR. Liberty Roofing and Gutter, Inc., Watertown, NY (2005-present). For this company which worked primarily on contracts at Fort Drum, inspected the work of 35 people on four crews.

SUPERINTENDENT & QUALITY CONTROL MANAGER. Upstate Construction, Watertown, NY (2001-05). Handled safety reports and inspections, daily reports, equipment inspections, preparatory meeting with all sub-contractors, security of job site and tools, daily police call of job site, inspections of all electrical cords, outlets and tools, problems on job sites, safety and welfare on job site, and the receiving and inventory of all materials on site.

QUALITY CONTROL/SAFETY SUPERVISOR. Fort Drum, NY.
2000: Fire Station Exhaust System. Handled installation for all fire stations on this major Army base and its adjacent training base.
1999: 10th Mountain Battle Simulation Center. Installation of metal building and vault.

COLLEGE STUDENT. Jefferson Community College, Watertown, NY (1993-98). While involved in horticulture, cabinetmaking, and carpentry programs of study, excelled in numerous summer jobs for landscaping companies.

PLATOON SERGEANT & TECHNICAL INSPECTOR. U.S. Army, Fort Bragg, NC (1990-93). Oversaw the welfare of 15 individuals as well as their training while also supervising the safe utilization for 15 helicopters; scheduled flight crews, performed inspections, handled maintenance inventories, and had vast responsibility for both human and material resources. Also functioned as a Technical Inspector Orders (T.I. Orders).

TECHNICAL INSPECTOR. U.S. Army, Italy (1988-90). Inspected aircraft, inspected work of enlisted men, maintained oil analysis charts, performed safety inspections of hangars, taught safety classes, and was in charge of all historical/log book of all aircraft parts.

CREW CHIEF & TECHNICAL INSPECTOR. U.S. Army, Fort Drum, NY (1984-88). As a Crew Chief, was responsible for all equipment on aircraft, all inspections, all maintenance in area, training, weapons, missions.

- As Technical Inspector, handled all inspections – scheduled and unscheduled; inspected safety procedures, log book, and supervised overall checking of aircraft and log books.

Other experience: DOOR GUNNER. U.S. Army, Korea (1982-83). Acted as a Door Gunner on the UH-1H Helicopter. Assured that all weapons were clean and operable and that aircraft was loaded with ammunition as well as food and water for extended missions.

PERSONAL

Can provide outstanding personal and professional references.

Accident prevention programs

Date

Exact Name of Person
Title or Position
Name of Company
Address
City, State, Zip

SAFETY OFFICER

Dear Exact Name of Person: (or Sir or Madam if answering a blind ad.)

With the enclosed resume, I would like to introduce myself and the extensive experience I offer in the field of safety management.

As you will see from my resume, I have designed safety standards, codes, and safety requirements for Occupational Safety and Health (OSHA) compliance; DOD safety and health standards; and hazardous waste materials handling and training procedures. I am considered a leading expert in aviation safety including air traffic control, refueling operations, aeromedical evacuations, airport security and anti-terrorism, fire prevention, and many other areas. Skilled in implementing Total Quality Management concepts in all aspects of safety program management, I am an experienced auditor of safety programs with the goal of reducing hazards in the workplace.

In my most recent position, I excelled as Safety Officer and senior safety expert. I utilized my knowledge and personal initiative to make changes which reduced accidents, improved risk assessment, and refined hazard awareness plans while authoring numerous procedures, policies, and regulations designed to safeguard lives and property. I have excelled as a Director of Safety supervising a 20-person staff in an organization with 400 employees, 125 vehicles, and 65 aircraft.

In prior jobs I have managed safety programs for medical organizations, and I have served as a medical evacuation pilot both during combat and in peacetime. I once served as an Aeromedical Evacuation Pilot in an air ambulance company in Germany. I began my military service as a pilot and, prior to advancing to the safety management field, was a UH-60 and UH-11 Rotary Wing Pilot.

I believe my hands-on experience as a pilot has given me an advantage over my safety counterparts, since I became accustomed early in my career to operating with the attitude that there is "no room for error." As a Safety Director, I have become respected for my enthusiastic and thorough approach to safety management, and I proud that I have helped every organization in which I have worked achieve new levels of safety accomplishment. Where safety is concerned, there really is no room for error.

If you can use an accomplished safety professional who can provide outstanding personal and professional references, I hope you will contact me to suggest a time when we might meet to discuss your needs and goals and how I might help you achieve them. Thank you in advance for your time.

Sincerely,

Silas Calvin

SILAS CALVIN

1110½ Hay Street, Fayetteville, NC 28305 • preppub@aol.com • (910) 483-6611

OBJECTIVE

To benefit an organization that can use an experienced safety management professional with expertise related to accident investigation, risk identification and control, hazardous materials and hazardous waste disposal, loss control, as well as safety program development.

SAFETY KNOWLEDGE

Offer a vast range of technical knowledge, skills, and experience including the following:

- Have designed safety standards, codes, and safety requirements for Occupational Safety and Health (OSHA) compliance; DOD safety and health standards; hazardous waste materials handling and training.
- Considered a leading expert in transportation and aviation safety including air traffic control, refueling operations, aeromedical evacuations, airport security and anti-terrorism, fire prevention, and all other areas.
- Skilled in implementing Total Quality Management concepts in all aspects of industrial safety program management. Experienced auditor of safety programs to reduce hazards.

EXPERIENCE

SAFETY OFFICER. U.S. Army, Italy and Fort Drum, NY (2004-present). As a Chief Warrant Officer 4, was the senior safety expert within a 400-person organization; dramatically reduced accidents, improved risk assessment, developed hazard awareness plans, and authored numerous procedures, policies, and regulations designed to safeguard both lives and property.

- Guided, trained, and supervised 20 junior safety officers in developing accident prevention programs. Conducted safety surveys on landing zones, parking zones, and field site aircraft; established safety procedures and ensured safe aircraft parking areas day and night.
- Checked air traffic control procedures, ensured aircraft refueling operations were performed safely, and continuously conducted risk assessments.
- Was credited with being the driving force behind ensuring that safety awareness was understood and practiced by all employees; raised safety consciousness to all-time-high.
- Developed and managed an aviation safety program involved in operations ranging from tactical air assault to VIP transport, and from administrative airlift in Italy to military support activities in Iraq, Afghanistan, Kuwait and Pakistan.

SAFETY OFFICER. U.S. Army, Fort Drum, NY (2003-04). At the busiest U.S. military base, oversaw all matters related to aviation safety which included providing oversight for two airfields as well as for seven aviation organizations with more than 350 aircraft.

- Implemented safety standards, codes, and safety requirements for OSHA, DOD, and hazardous waste materials handling and training.

SAFETY OFFICER. U.S. Army, Fort Drum, NY (1999-03). Directed the safety program for the 42d Aviation Brigade, a large and complex organization; improved safety in ground and aviation operations. In an Aeromedical Evacuation Detachment, oversaw the safety program affecting 70 personnel and 7 UH-60 Black Hawk helicopters; instituted an innovative new four-step method of developing, implementing, and monitoring the safety program.

EVACUATION PILOT. U.S. Army, Germany (1998-99). In Germany, initiated a complete review of airfield policies and procedures which uncovered shortcomings which we corrected. Developed and implemented an effective safety awareness/accident prevention program.

EDUCATION & TRAINING

Received **Associate of Arts degree**, Northwest Technical College, Bemidji, MN, 1983.
Honor graduate of the **Air Force Aircrew Life Support Officer Course**, 1990.
Graduated from the **Army Aviation Safety Course**, Fort Bragg, NC, 1995.

Accident investigation and prevention

Date

Exact Name of Person
Title or Position
Name of Company
Address
City, State, Zip

SAFETY OFFICER

Dear Exact Name of Person: (or Dear Sir or Madam if answering a blind ad.)

With the enclosed resume, I would like to make you aware of my interest in the position as Safety Engineer.

As you will see from my enclosed resume, I have excelled in a track record of accomplishment as an Aviation Safety Officer with the U.S. Army. Because of my safety expertise, I have earned promotion to the rank of Chief Warrant Officer 4 and have been recommended for promotion to CW5. I have decided, however, that I would like to retire from the U.S. Army and embark upon a career in aviation safety with a civilian organization.

I began my career in aviation safety as an Aviator and then performed with distinction in jobs as an Aircraft Maintenance Technician, Maintenance Test Pilot, and—more recently—Aviation Safety Officer. In my current position as Aviation Safety Officer at the nation's U.S. military base, I coordinate programs related to work safety, aviation product safety, accident prevention, as well as compliance with local, state, and federal regulations. I have developed accident-prevention safety programs related to aviation maintenance, hazardous material storage and handling, as well as radiation commodity storage and handling. In my previous position as Aviation Safety Officer in Germany, I reduced ground accidents by 50% while organizing pre-accident plans, writing safety standard operating procedures, and implementing environmental clean-up operations. In a previous position, I created a new radiation protection program for nondestructive X-ray equipment which became a model for the parent organization. I also transformed a hangar described as a "hazardous material disaster" into a facility exhibiting a high degree of environmental awareness.

Considered one of the nation's leading aviation safety authorities, I have served as an official U.S. Army Accident Investigation Board member. I have played a key role in consulting and problem solving in numerous mishaps including a helicopter mishap in Alabama in 2003, an Apache mishap in Korea in 2000, and an Apache mishap in 1999. Fluent in Italian and proficient in German, I have worked with aviation officials all over the world as well as with the officials of major companies including Boeing and Lockheed Martin.

You will notice from my resume that, in addition to my extensive training and certifications, I hold two master's degrees: one in Aviation/Aerospace Safety and the other in Airline Operations. I am held in the highest regard in the international aviation community and can provide outstanding references at the appropriate time.

Sincerely,
Rick Sanford

RICK SANFORD

1110½ Hay Street, Fayetteville, NC 28305 • preppub@aol.com • (910) 483-6611

OBJECTIVE

To contribute to an organization that can use a highly skilled aviation safety professional who offers experience in providing analysis and recommendations related to aircraft damage prevention strategies, developing safety programs, conducting investigations of mishaps, analyzing trends, and preparing economic and financial reports.

EDUCATION

Master of Aeronautical Science in Aviation/Aerospace Safety, Troy State University in Dothan, AL, 2005.
Master of Aeronautical Science in Airline Operations, Troy State University, 2005.
B.S. in Biology, minor in Chemistry, Sierra College, Rocklin, CA, 1982.

EXPERIENCE

Have excelled in a track record of promotion to Chief Warrant Officer 4 (CW4) while becoming one of the military's foremost safety experts, U.S. Army:

SAFETY OFFICER. Fort Rucker, AL (2005-present). At the nation's U.S. military base, manage ground and aviation safety for 650 individuals and 240 aircraft; coordinate programs related to worker safety, aviation product safety, accident prevention, as well as compliance with local, state, and federal regulations.

- Developed numerous accident-prevention safety programs including programs related to aviation maintenance, hazardous material storage and handling, as well as radiation commodity storage and handling.
- Provided expert testimony and safety engineering case management in aviation accident and mishap investigation; I am skilled at accident reconstruction and analysis to determine cause and suggest solutions.

SAFETY OFFICER. Germany (2004-05). Planned and organized a safety program for an organization with 800 individuals, 270 aircraft, 27 major customers, and 86 million dollars.

- Planned and organized a pre-accident plan, safety SOP, and conducted monthly walk-through inspections of all hangars, workplaces, shops, motorpools, and field sites.
- Reduced military and civilian ground accidents and sports injuries by 50%. Planned and implemented an environmental clean-up operation for a central receiving and storage area for hazardous materials. Performed severe weather/disaster relief rehearsals.

QUALITY CONTROL LEADER & AVIATION SAFETY OFFICER. Fort Rucker, AL (1995-03). For two separate organizations with hundreds of employees, managed the safety program, hazardous materials program, radiation safety program, and quality assurance.

- On my own initiative, created and implemented a new radiation protection program for nondestructive X-ray equipment which became a model for the parent organization.
- Through my hazardous materials knowledge, transformed a hangar described as a "hazardous material disaster" into a facility exhibiting a high degree of environmental awareness. Coordinated the removal of underground waste oil containers and set up a drum containment system which met EPA requirements.
- Developed a hazardous waste program "from scratch."

HIGHLIGHTS OF AVIATION ACCIDENT EXPERTISE

Have served as an official U.S. Army Accident Investigation Board member.
2003: Provided expert safety consulting related to a Blackhawk helicopter mishap, Fort Rucker. Assisted in investigating cause of accident and training requirements for crew.
2000: Provided safety leadership during an Apache mishap in Korea. Investigated cause, secured wreckage, estimated damage, coordinated with Italian officials.
1999: Maintenance Officer in charge of Estimate Cost of Damage (ECOD) assessment in an Apache mishap; worked with Boeing and Lockheed Martin.

Safety program development

Date

Exact Name of Person
Title or Position
Name of Company
Address
City, State, Zip

SAFETY OFFICER

Dear Exact Name of Person: (or Dear Sir or Madam if answering a blind ad.)

I would appreciate an opportunity to talk with you soon about how I could contribute to your organization through my managerial talents and my specialized experience in safety program development and administration.

Throughout my distinguished career as a U.S. Army warrant officer, I have been consistently described as a true professional who sets the standard. I offer B.S. and A.S. degrees in Professional Aeronautics with an area of concentration in aviation safety along with more than 5,000 hours of flight time. My training has included advanced programs in crash survival investigation, instructor pilot procedures and techniques, and the U.S. Army Aviation Safety Officers Course, as well as several advanced management courses.

My present position involves overseeing aviation safety activities for a 1,250-person organization as the senior advisor to the chief executive. In a short period with this organization I have been credited with making significant improvements in the quality of the program.

Known as a mature and dependable leader and manager, I also offer a reputation as an enthusiastic and highly energetic individual with a creative mind and the ability to think on my feet. I am highly skilled at motivating and encouraging others to perform up to my high standards.

I hope you will call or write me soon to suggest a time convenient for us to meet and discuss your current and future needs and how I might serve them. Thank you in advance for your time.

Sincerely yours,

Phillip Wilbur

PHILLIP WILBUR

1110½ Hay Street, Fayetteville, NC 28305 • preppub@aol.com • (910) 483-6611

OBJECTIVE

To offer my expertise in the field of safety coordination and program administration along with my general management experience to an organization that can use a strong leader who has refined natural abilities in a distinguished career as a military warrant officer.

EDUCATION & TRAINING

B.S. and A.S. degrees in Professional Aeronautics, Central Texas University with a concentration in aviation safety.
Excelled in numerous advanced training programs in aviation safety, management, crash survival investigation, flight training, and techniques for instructor pilots.

EXPERIENCE

SAFETY OFFICER. U.S. Army, Fort Drum, NY (2005-present). Serve as the resident expert on aviation safety matters and chief advisor to a senior executive in a 1,250-person organization with more than 120 aircraft valued at almost $386 million.

- Officially described as a "superb trainer, leader, and mentor," was credited with significantly improving the safety consciousness and quality of programs for the organization.

Advanced in the field of aviation management, U.S. Army, Fort Bragg, NC:
AVIATION SAFETY MANAGER. (2003-05). Handpicked on the basis of my expertise and performance, and was chosen to advise the senior executive of a unique organization with more than 250 aviation professionals on information pertaining to safety issues.

- Determined that a problem existed concerning a railroad crossing and the landing gear on one type of aircraft, made recommendations, and briefed officials on solutions.
- Cited as "extremely talented and knowledgeable," was highly successful in integrating safety guidelines into all aspects of training and operations.
- Was personally credited with setting the safety standards which led the organization to far exceed government standards and significantly lower accident rates.

INSTRUCTOR PILOT. (2002-03). Provided counseling, supervision, instruction, and training for students at the Army's national aviation training center.

- Highly successful in guiding and relating to young pilots, producing top-notch students who maintained an impressive 100% pass rate in the rigorous and demanding program.
- Flew in excess of 1,000 hours while training more than 2,200 students.
- Refined my communication skills while presenting daily briefings to top-level executives.

TRAINING PROGRAM MANAGER. (2000-02). Supervised ten specialists in a department which provided training, advice, and counseling for 500 aviation management trainees; maintained personnel records and ensured the quality of training.

AVIATION PERSONNEL MANAGER. U.S. Army, Fort Wainwright, AK (1997-99). Ensured that multinational aviators were properly assigned to organizations throughout the country by maintaining their skills and performance records.

SAFETY PROGRAM COORDINATOR. U.S. Army, Fort Rucker, AL (1993-97). Led personnel at the National Training Center to a zero accident rate even during grueling and dangerous desert training while working closely with both civilian and military agencies to improve and develop safety programs.

AIRCRAFT EXPERTISE

C-150, 172, and 172 RG & 210; M-21 (T-41A/182; UH-1 A, B, C, D, H, M, V, and V Bell 204 and 205 series; OH-58 A and C Bell 206 Series; UH-60 A/L (S-70 Sikorsky).

Municipal sanitation services

Date

Exact Name of Person
Title or Position
Name of Company
Address
City, State, Zip

SANITATION QUALITY SUPERINTENDENT

Dear Exact Name of Person: (or Dear Sir or Madam if answering a blind ad.)

With the enclosed resume, I would like to make you aware of my interest in exploring employment opportunities with your organization.

In my current position as a Sanitation Quality Superintendent, I manage approximately 130 employees operating a fleet of 45 vehicles used in solid waste collection and disposal for a progressive and growing city with a current population of over 113,000. Through my leadership, we have achieved an 87.9% customer satisfaction rate in a survey of city residents, a rate of satisfaction almost unheard of in any city; and we have reached this high level despite the drawbacks of using outdated equipment.

You will notice from my resume that I "worked my way up" to my current position after excelling in entry-level jobs as an Equipment Operator and Sanitation Route Inspector.

I hope you will call or write me soon to suggest a time convenient for us to meet to discuss your current and future needs. Thank you in advance for your time.

Sincerely yours,

Stefan Woodlock

Alternate last paragraph:
I hope you will welcome my call soon to arrange a brief meeting when we might meet to discuss your needs and goals and how my background might serve them. I can provide outstanding references at the appropriate time.

STEFAN WOODLOCK

1110½ Hay Street, Fayetteville, NC 28305 • preppub@aol.com • (910) 483-6611

OBJECTIVE

To offer a background of progression in managerial roles based on outstanding analytical skills and sensitivity to customer comments and needs as well as extensive knowledge of solid waste operations to an organization in need of a mature professional.

EXPERIENCE

Advanced in this track record of accomplishments with the City of Kenai, AK, while becoming known as a professional who could build teams and make the most of outdated and inadequate equipment:

SANITATION QUALITY SUPERINTENDENT. (2005-present). Manage approximately 130 employees operating a fleet of 45 vehicles used in solid waste collection and disposal for this progressive and growing city with a current population of over 113,000.

- Achieved an 87.9% customer satisfaction rate in a survey of city residents, a rate of satisfaction almost unheard of in any city; reached this high level despite the drawbacks of using outdated equipment.
- Earned frequent praise for my ability to listen to customer concerns and questions and find the solutions to difficult and sensitive issues.
- Planned, directed, and coordinated the department's work flow including scheduling, reviewing and evaluating work performance areas, and working with staff members to solve problems.
- Made hiring decisions; trained and evaluated employee performance; handled disciplinary actions and terminations. Participated in budget development and administration including forecasting future needs and approving expenditures.
- Was commended for my sound judgment and superb understanding of how to move equipment and personnel around in order to meet the greatest needs and for overcoming such handicaps as backups and delays caused by renovation at one landfill.

SANITATION ROUTE INSPECTOR. (2001-05). Was promoted to the superintendent's position on the basis of my professionalism and accomplishments while supervising 14 employees responding to meeting the solid waste collection needs of city residents.

- Managed a fleet of body trucks and one Knuckle boom in a sanitation operation.
- Known for my dependability, helped ensure that my route personnel maintained a reputation with the public for always being on schedule and providing efficient service.
- Selected to serve on a merit pay review committee, made suggestions based on my concerns, which were investigated and resulted in more equitable distributions.

MEDIUM EQUIPMENT OPERATOR. (2000-01). Joined this organization with no prior knowledge or experience in solid waste collecting and quickly became known for my hard work and willingness to learn while providing city residents with timely service.

- Operated a truck with two crew members following a scheduled route within the city and learned the different types of vehicles and equipment used in solid waste collection.

TRAINING

Completed several City of Kenai-sponsored seminars and training programs emphasizing such topics as employee/management relations, customer relations and service, and how to prevent abuse of leave policies.

Attended a two-day program in residential route design for better service sponsored by the City of Kenai Waterworks; and a 40-hour management development course, Alaska Vocational Technical Center, Seward, AK.

PERSONAL

Excel in providing leadership by example: make it personal policy to never expect any employee to do anything I am not willing to do myself. Thrive on challenge and pressure.

OEM manufacturing

Date

Exact Name of Person
Title or Position
Name of Company
Address
City, State, Zip

SENIOR QUALITY ENGINEER

Dear Exact Name of Person: (or Dear Sir or Madam if answering a blind ad.)

With the enclosed resume, I would like to acquaint you with my exceptional skills and years of experience as an industrial and quality engineer with a solid background in manufacturing, quality assurance, and project management.

As you will see from my resume, I have worked at the same facility since before Allied Industries took over the operation from Monarch, Inc. In my years at this plant, my loyalty to the company and outstanding problem-solving skills, strong personal initiative, and extensive knowledge of all phases of industrial, manufacturing, and quality engineering have allowed me to progress into positions of increasing responsibility.

As Senior Quality Engineer, I have been responsible for increasing first-pass yield of a complex manual assembly from 3% to 56%, exceeding the company objective six months before the projected deadline to meet that goal. I served on the Certification Committee that achieved ISO 9002 certification for the facility, and I have worked hard to ensure increased productivity by increasing awareness of initial quality, reducing rework and warranty cost by more than 50%.

I have earned a Master of Business Administration degree and also possess a Bachelor of Science in Electrical Engineering. I have also supplemented my degree programs with graduate-level courses on planning, scheduling, and inventory control.

If your organization could benefit from the services of a talented and self-motivated industrial, electrical, or manufacturing engineer, I hope you will contact me. I assure you that I have an excellent reputation as a loyal and dedicated worker, and would quickly become a strong asset to your operation.

Sincerely,

Irving B. Dellon

IRVING B. DELLON

1110½ Hay Street, Fayetteville, NC 28305 • preppub@aol.com • (910) 483-6611

OBJECTIVE

To benefit an organization that can use an experienced, self-motivated, and educated quality and industrial engineer with and a background in manufacturing and project management.

EDUCATION

MBA in General Management, Kansas State University, Lawrence, KS, 1997.
Bachelor of Science in Electrical Engineering, 1984.
Graduate-level course on Production planning, scheduling, and inventory control, Kansas State University, Lawrence, KS, 1989.

EXPERIENCE

SENIOR QUALITY & MANUFACTURING ENGINEER. Allied Industries (formerly Monarch, Inc.), Lawrence, KS (1990-present). Started with Monarch, Inc. as a Manufacturing Engineer; became Senior Manufacturing Engineer, a job title that was changed to Senior Quality Engineer after Allied Industries took over the plant in 1995.

- Interacted with Original Equipment Manufacturer (OEM) customers, manufacturing and production departments to deliver high-quality products according to customer specifications.
- Increased first-pass yield of a complex manual assembly and wiring operation from 3% to 56% in a 15-month period. Company objective was to achieve 55% by the end of 2001; achieved this goal by June of 2000.
- Raised productivity by increasing quality awareness; reduced rework and warranty cost by more than 50%.
- Served on the plant's ISO Certification Committee; facility was awarded ISO 9002 certification.
- Created a new layout for the production line to reduce inventory and transfer the manufacturing process into a "just in time" process.
- Completed hazardous waste management projects and waste minimization projects to comply with EPA, SARA, and OSHA regulations; coordinated UL activities and procedures.

Quality Control: *Used the following quality control principles in my position as Quality Engineer:*

- Apply statistical process control in fabrication, assembly and finishing operations. Design and process FMEA. Quality audits of products and process. Establish and monitor quality control programs. Team leader and facilitator of corrective action and continuous cycle time improvement teams. Plan and implement ISO 9001 and 9002 quality system, periodic audit, and document control. Implement OEM specific Customer requirements.

Manufacturing/Industrial Engineering: *Used the following manufacturing and industrial engineering processes and principles in my position as a Manufacturing/Industrial Engineer.*

- JIT system of manufacturing, which includes:
  Standardization of designs, processes, and equipment — Group technology
  Demand flow technology in the assembly process — KANBAN inventory control
  Powder paint system evaluation and implementation — CIM/Cell Technology

General Manufacturing/Industrial: *Apply the following techniques and principles:*

- Implement new products and designs into manufacturing
- Resolve manufacturing problems
- Provide technical support in assembly and fabrication area
- Run computer simulations of new methods and layouts
- Implemented new copper manufacturing cell and welding booth

PERSONAL

Outstanding personal and professional references are available upon request.

Hazardous chemical disposal

Date

Exact Name of Person
Title or Position
Name of Company
Address
City, State, Zip

SENIOR QUALITY ENGINEER

Dear Exact Name of Person: (or Dear Sir or Madam if answering a blind ad.)

I would appreciate an opportunity to talk with you soon about how I could contribute to your organization through my extensive background in quality control and engineering. I can provide outstanding personal and professional references upon your request, and I have won numerous awards for my resourceful problem solving as well as for my excellent work habits including perfect attendance.

As you will see from my resume, I have most recently made valuable contributions to the Inova Corporation and, on my most recent annual performance evaluation, I received the highest rating given on every area of performance measured. While in this job I figured out, on my own initiative, a resourceful and low-cost method of disposing of hazardous chemical waste which is saving the company thousands of dollars annually. I have also developed a system of desk-top procedures which has become the model for similar Inova Corporation sites. These 20 procedures are now being used in the U.S. and overseas and assure that field site depots are in audit-ready condition at all times. These and other procedures I have developed have led to more consistency on the job and far fewer errors. My job knowledge, creativity, problem-solving ability, and decision-making skills are at an exceptionally high level, and I feel I could make valuable contributions to your organization, too.

You will also see from my resume that I previously "cut my teeth" on quality control inspection and testing while working at the GE Corporation. Even during college, while earning my degree in Business Administration, I worked as a quality control inspector for companies in the Ohio area.

You would find me to be a congenial professional who is known for my ability to develop and maintain excellent working relationships. In my current job I have been credited with greatly improving internal communication and trust through my tactful communication skills and gracious style of dealing with people

I hope you will write or call me soon to suggest a time when we might meet to discuss your current or future needs and how I might serve them. Thank you in advance for your time.

Sincerely yours,

Leonard Jackson

LEONARD JACKSON

1110½ Hay Street, Fayetteville, NC 28305 • preppub@aol.com • (910) 483-6611

OBJECTIVE

To apply my extensive knowledge related to quality control and engineering to an organization that can use a self-motivated self-starter and team player with outstanding communication skills who offers a proven ability to develop new methods and reduce costs.

EXPERIENCE

SENIOR QUALITY ENGINEER. Inova Corporation, Inc., Various locations (2000-present). Have excelled in versatile assignments which tested my ability to articulate technical concepts and solve stubborn problems related to safety and quality:

Hopkinsville/Fort Campbell, KY (2005-present). At one of the world's busiest U.S. military base, have received the highest possible rating on every area of my performance measured: *initiative* ("self starter"); *follow-through* ("never any loose ends"); *interpersonal relations* ("outstanding rapport with management and engineering"); *problem solving* ("solves problems with minimum direction"); *team work* ("a team player"); and *job knowledge* ("performs tasks that are beyond scope").

- Supervise the receipt, in-process, and the final and shipping inspections of the Target Acquisition Designation Sight/Pilot Night Vision Sensor (TADS/PNVS) flight hardware related to the Apache helicopter; became an expert in applying MIL-I-45208A in inspections, MIL-STD-45662 in calibrating electronic test equipment, and MIL-Q-9858A in processing flight hardware.
- On my own initiative, aggressively investigated hazardous chemical disposal options and devised a new method for disposing of expired chemicals/adhesives/paints that reduced annual costs from $10,000 to $200 at each field site; this led to Desk Top SOP-016 now being implemented at all Inova Corporation sites.
- Developed a thoroughly documented system of 20 procedures for field site depots to assure audit-ready condition at all times; these procedures have become the model for all TADS sites and are being used by other quality representatives at Inova depots throughout the U.S. and oversees; have been praised in writing for ensuring "more consistency on the job and far fewer errors."
- Developed and implemented an inspection system that ensures outstanding performance at field sites; this is a cost-effective system that exceeds both contractual and internal requirements. Designed and established a calibration system praised as comprehensive.

Iraq (2004). During the War or Terror, played a major role in operating a quality control center supporting Apache helicopters. Worked unusually long hours seven days a week over a 43-day period to produce exceptionally high mission capability rate under harsh conditions.

Watertown/Fort Drum, NY (2003). As the sole quality expert, made sound decisions and set standards while conducting inspections upon receipt, during processing, and before shipping approximately 9,084 pieces of hardware annually.

Boston, MA (2000-02). Learned processes specific to electro-optical targeting/night vision equipment using engineering drawings, military specifications, and contracts. Established inspection criteria for 300 manufacturing process plans.

QUALITY CONTROL INSPECTOR/TESTER. GE Corporation, Toledo, OH (1986-99). Received awards for perfect attendance four years in a row and was promoted three times while supervising final inspection before shipment to include precise electronic and mechanical checks, adjustments, and functional tests for copiers, duplicators, and laser fax machines.

EDUCATION

B.A. degree, Business Administration, University of Massachusetts, Boston, MA, 2001.
A.A. degree, Business Administration, Owens Community College, Toledo, OH, 1991.

PERSONAL

Member, American Society for Quality Control. Outstanding references available.

Commercial freight services

Date

Exact Name of Person
Title or Position
Name of Company
Address
City, State, Zip

SITE FREIGHT COORDINATOR

Dear Exact Name of Person: (or Dear Sir or Madam if answering a blind ad.)

I would appreciate an opportunity to talk with you soon about how I could contribute to your organization through my experience in all aspects of traffic and transportation management. I offer extensive knowledge of LTL, TL, Intermodal, rate negotiations, pool shipments, and cost analysis to determine the most economical method of shipping.

As you will see from my resume, I am currently site freight coordinator for a Fortune 500 company, and I have continuously found new ways to reduce costs and improve efficiency while managing all inbound and outbound shipping. On my own initiative, I have recovered $10,000 in claims annually while saving the company at least 40% of a $10 million LTL budget. In addition to continuous cost cutting, I have installed a new bar code system in the finished goods shipping area and have installed a new wrapping system.

In previous jobs supervising terminal operations, I opened up new terminals, closed down existing operations which were unprofitable, and gained hands-on experience in increasing efficiency in every terminal area.

With a reputation as a savvy negotiator, I can provide excellent personal and professional references. I am held in high regard by my current employer.

I hope you will call or write me soon to suggest a time convenient for us to meet and discuss your current and future needs and how I might serve them. Thank you in advance for you time.

Sincerely yours,

Pedro Palacios

Alternate last paragraph:
I hope you will welcome my call soon to arrange a brief meeting at your convenience to discuss your current and future needs and how I might serve them. Thank you in advance for your time.

PEDRO PALACIOS

1110½ Hay Street, Fayetteville, NC 28305 • preppub@aol.com • (910) 483-6611

OBJECTIVE

To contribute to an organization that can use a skilled traffic management professional who offers a proven ability to reduce costs, install new systems, optimize scheduling, negotiate rates, anticipate difficulties, solve problems, and keep customers happy.

EXPERIENCE

SITE FREIGHT COORDINATOR. DuPont Corporation, Wilmington, DE (2000-present). For this Fortune 500 company, have continuously found new ways to cut costs and improve service while managing all inbound transportation as well as outbound shipping totaling in excess of one million dollars in finished goods daily; supervise ten people.

- Saved the company at least 40% of a $10 million LTL budget by resourcefully combining my technical knowledge with my creative cost-cutting skills.
- Recovered $10,000 annually in claims; prepare all cargo claims documents for corporate office and oversee all procedures for proper claims documentation.
- Installed a bar code system in Finished Goods Shipping, and also installed a new wrapping system.
- Reduced overtime by 90% while simultaneously cross-training some employees and improving overall morale.
- Became familiar with Total Quality Processes while analyzing transit times to ensure consistent and timely Just-In-Time delivery schedules.
- Am a member of the B & D corporate committee for North American rate negotiations; negotiate rates with various carriers on special moves.
- Justify capital appropriation requests for funding special projects; audit all freight bills and process them for payment.
- Prepare all documents for export shipments to Canada; also advise about the shipment of hazardous materials and maintain proper documentation placards and labels.
- Coordinate all site printing of product information and warranty cards.
- Am responsible for site switcher and equipment such as leased trailers.
- Have earned a reputation as a savvy negotiator with an ability to predict future variables that will affect traffic costs.

SUPERVISOR. International Freightways, Inc., Atlanta, GA (1993-00). Supervised up to 12 drivers while managing second-shift operations and controlling inbound and outbound freight at this terminal operation.

- Increased efficiency in every operational area; improved the load factor, reduced dock hours, and ensured more timely deliveries.

INVENTORY SPECIALIST. La-Z-Boy East, Inc., Florence, SC (1989-92). Learned the assembly process of this name-brand furniture manufacturer while managing replenishment of subassemblies for daily production.

Highlights of other experience:

- As Terminal Manager for Spartan Express, opened a new terminal in South Carolina; determined the pricing structure, handled sales, and then managed this new operation which enjoyed rapid growth.
- Gained experience in closing down a terminal determined to be in a poor location.
- As Operations Manager for a break bulk operation, supervised up to 12 people in a dock center while managing the sorting/segregating of shipments from origin to destination.

EDUCATION

Studied business management and liberal arts, Ohio State and LaSalle University. Completed extensive executive development courses in the field of transportation and traffic management sponsored by University of Toledo and Texas Technical University

PERSONAL

Can provide outstanding personal and professional references. Will relocate.

Maintenance and repair contracts

Date

Exact Name of Person
Title or Position
Name of Company
Address
City, State, Zip

SITE SAFETY MANAGER

Dear Exact Name of Person: (or Dear Sir or Madam if answering a blind ad.)

With the enclosed resume, I would like to make you aware of my background in computer networking and satellite communications system operations as well as of my reputation as a methodical professional with strong troubleshooting skills.

As you will see from my resume, I have advanced to hold multiple simultaneous roles with Cavuto International Corporation, a contracting firm which provides maintenance and repair support for the state-of-the-art systems used to provide signal support for the 101st Airborne Division based in Fort Campbell, KY. Originally hired as a Senior Computer Systems Specialist for this firm, I advanced to Senior Contract Representative and now serve as Site Safety Manager for 18 customer units with multiple systems in diverse locations in Kentucky, North Carolina, New York, and Alaska. I have traveled to locations including Iraq and Afghanistan to provide technical assistance. In my current job, I am heavily involved in Quality Assurance, and I develop all checklists, protocols, and tests for QA.

I have been given opportunities to work with the most sophisticated computer networking and signals communication systems used by the U.S. military to include LANs, Mobile Subscriber Equipment (MSE), and satellite communications systems. Widely known for my expertise as a troubleshooter, I have managed projects to upgrade 40 Cisco routers to ensure their compliance with yearly requirements and another to fabricate cable assemblies in a 2004 project which came in on schedule.

Earlier while serving in the U.S. Army as a technical instructor and electronics technician, I advanced to supervisory jobs and earned honors including two U.S. Army Commendation Medals for my leadership, dedication, and technical expertise. With a Top Secret clearance and reputation for high personal standards of integrity and honesty, I have excelled in extensive technical in addition to completing college studies in Electrical Installation and Maintenance, Computer Science, Electrical Engineering, and electronics Technology.

If you can use a skilled technician with expertise in computer networking and satellite communications, I hope you will contact me to suggest a time when we might meet briefly to discuss your goals and how my background might serve your needs. I can provide outstanding references at the appropriate time.

Sincerely,

Sheridan Hayes

SHERIDAN HAYES

1110½ Hay Street, Fayetteville, NC 28305 • preppub@aol.com • (910) 483-6611

OBJECTIVE

To offer my considerable technical expertise to an organization that can benefit from my knowledge of computer networking and satellite communications, skills in troubleshooting, and reputation for being logical and methodical with excellent interpersonal skills.

EDUCATION

Diploma in **Electrical Installation and Maintenance**, Hopkinsville Community College, KY, 2001.
Completed additional studies in **Computer Science, Electrical Engineering, and Electronics Technology** at the college level.

EXPERIENCE

SITE SAFETY MANAGER (2004-present); SENIOR CONTRACT REPRESENTATIVE & SENIOR COMPUTER SYSTEMS SPECIALIST (2000-04). Cavuto International Corporation, Fort Campbell, KY (2000-present). Have advanced to management roles with this contracting firm which provides maintenance and repair support for satellite communications, KU and C band equipment, and LANs utilized by the 101st Airborne Division to meet their worldwide commitment to responding on short notice anywhere in the world.

- As **Site Safety Manager from 2004-present,** am currently accountable for over $10 million in government property on a multimillion-dollar contract, and my efforts played a key role in netting a cost avoidance of $4 million for the 101st Airborne Division. Handle all modifications and work improvements of a multimillion-dollar equipment inventory.
- Handled extensive quality assurance functions including the development of all checklists, tests, and protocols.
- Current provide outstanding customer support to multiple customer organizations, each of which has multiple systems; am supporting systems at Fort Campbell, Fort Bragg, and Fort Drum, and Fort Richardson and all these systems deploy to numerous other sites worldwide. Also work with Mobile Subscribe Equipment (MSE) and am proficient with test and diagnostic equipment.
- Utilize my extensive satellite communications background to troubleshoot all links; additionally troubleshoot computer networks.
- Have traveled to many countries including Iraq and Afghanistan to provide the technical assistance which ensures error-free and uninterrupted communications and computer networking support for the military's tactical mobile satellite systems.
- Work with Cisco routers on a regular basis and recently completed a project in which 40 routers had yearly upgrades installed into their UNIX software.
- Have been cited for my skill in managing a ten-person staff which includes seven electronics/computer technicians and three office support personnel; the technicians provide expertise related to pneumatics, electronics, electrical systems, hydraulics, avionics, and environmental control units.

SENIOR FIELD SERVICES TECHNICIAN. Engineering Resources, Inc., Fort Riley, KS (1998-00). Became familiar with the operation of and support services provided by a contractor supporting U.S. Army operations on receiver/jamming systems, direction finding sets, and airborne surveillance systems. Became highly skilled in troubleshooting and aligning power generation equipment, environmental control systems, hydraulics, and system electronics.

CLEARANCE

Was entrusted with a **Top Secret** security clearance.

PERSONAL

Outstanding personal and professional references are available on request.

Human services programs

Date

Exact Name of Person
Title or Position
Name of Company
Address
City, State, Zip

STATE PROGRAM MANAGER, DEPARTMENT OF FAMILY SERVICES

Dear Exact Name of Person: (or Dear Sir or Madam if answering a blind ad.)

With the enclosed resume, I would like to make you aware of the background in program management and staff development which I could put to work for your organization.

While excelling in the track record of advancement which you will see summarized on my resume, I have applied my strong organizational skills in implementing new programs, organizing conferences and seminars, training and counseling professional-level employees, and transforming ideas into operational realities. On numerous occasions, I have developed effective formats for grant, reports, documents, and quality assurance protocol which have been described as "models."

In my current position, I have served as a Program Manager for the state of Maryland while spearheading the development of new housing options and employment opportunities for physically and mentally challenged adults. With a reputation as a vibrant and persuasive communicator, I routinely interface with legislators, state and federal officials, as well as with local program managers. It has often been my responsibility to take a new law and make sure it is efficiently and resourcefully implemented at the local level while assuring compliance with federal and state guidelines. I am continuously involved in teaching and training others–not only the professionals whom I directly supervise but also professionals regionally and locally who turn to me for advice and assistance in problem solving.

I feel confident that my resourceful leadership, expertise in staff training and staff development, and pragmatic approach to operations management and service operations delivery could be valuable to your organization. If you feel that you could use my considerable experience in initiating new programs, making existing programs work better, and establishing effective working relationships, I hope you will contact me to suggest a time when we might meet to discuss your needs and how I might serve them. I can provide outstanding personal and professional references at the appropriate time.

Yours sincerely,

Norah Stephens

NORAH STEPHENS

1110½ Hay Street, Fayetteville, NC 28305 • preppub@aol.com • (910) 483-6611

OBJECTIVE

To benefit an organization that can use a results-oriented program manager and skillful coordinator of services who offers a reputation for creative leadership, a proven ability to initiate and develop new programs, along with experience in administering existing services.

EXPERIENCE

STATE PROGRAM MANAGER. Maryland Department of Family Services, Baltimore, MD (2004-present). Supervise two professionals–a Residential Specialist and a Vocational Specialist–while establishing and administering statewide policies related to the provision of services and programs for the 3% of Baltimore residents afflicted with developmental disabilities.

- Have spearheaded the exciting development of new housing options for the physically handicapped (PH); whereas physically and mentally challenged individuals traditionally have had to live in group homes, I have provided leadership in pioneering a new program through which those individuals and families have purchased their own homes.
- Negotiated with a nonprofit credit union and the Maryland Housing Agency to develop a new mortgage product using the "Home Sweet Home" program as a model; supervised the development of literature explaining the product.
- Am considered an expert on the Maryland Portal Law which governs how handicapped individuals gain access to services provided by the state; travel extensively to local communities to work with mental health professionals and mental health centers to provide training and problem-solving assistance related to this law and other matters.
- Have worked tirelessly to assure maximum vocational opportunities for mentally and physically challenged individuals; played a key role on an advisory committee which explored ways to help high school graduates with developmental disabilities transition into jobs and training programs.
- Provided leadership during a statewide summit that increased employment options for disabled adults.

STATE TEAM LEADER. Maryland Division of Human Services, Baltimore, MD (2003-04). Supervised three professionals while traveling throughout the state to monitor statewide compliance with a Medicaid Program known as the Community Options Program which provides the developmentally disabled with housing options other than group homes.

- Coordinated statewide training for case managers and service providers.
- Was essentially in the role of Quality Assurance Manager as I developed review documents and quality control protocol to assure local compliance with state law.

REGIONAL PROGRAM COORDINATOR. O'Berry Center, Baltimore, MD (2001-03). Played a key role in increasing the efficiency of services provided to the developmentally disabled while acting as a Consultant in Maryland. Worked with state and federal regulators while functioning as the region's technical expert on the Community Options Program.

REGIONAL PROGRAM COORDINATOR. Anne Arundel County Area System, Annapolis, MD (1996-01). Supervised four professionals while also functioning as a Case Manager with a case load of 18 persons with developmental disabilities; administered 18 separate budgets.

Other experience: PROGRAM COORDINATOR. STAR, Inc., Baltimore, MD (1994-95). While earning my master's degree, oversaw operations of three group homes.

EDUCATION

Master of Arts in Education, University of Maryland, Baltimore, MD, 1996.
Bachelor of Science in Special Education, University of Baltimore, MD, 1992.

Non-Destructive Inspections

Date

Exact Name of Person
Title or Position
Name of Company
Address
City, State, Zip

SUPERVISORY QUALITY CONTROL INSPECTOR

Dear Exact Name of Person: (or Dear Sir or Madam if answering a blind ad.)

With the enclosed resume I would like to introduce you to a highly motivated and technically proficient professional who offers versatile experience in jobs requiring a high level of attention to detail and the ability to work independently while also overseeing team efforts.

My greatest strengths are my initiative, ability to handle pressure and deadlines, and capacity for dedicating my efforts to solve problems and exceed management expectations. Selected to receive extensive and advanced training, I have earned respect for my expertise in aircraft quality control and Non-Destructive Inspection (NDI) procedures. Having advanced to supervise five subordinates, I oversee support for A-10 and C-130 aircraft while employing the latest high-tech analytical methods to detect problems so they can be corrected and aircraft quickly returned to service.

My Air Force experience helped me develop strong skills in supply operations as well as in utilizing automated systems to maintain records and documentation. While working in these areas of operations, I have been credited with making significant contributions which reduced the time needed to procure critical parts as well as the downtime of aircraft because of delays in obtaining parts.

If you can use an adaptable self starter who will not give up until a problem is solved and a workable solution in place, please call or write me soon to suggest a time when we might have a brief discussion of how I could contribute to your organization. I will provide excellent professional and personal references at the appropriate time.

Sincerely,

Amanda Sullivan

AMANDA SULLIVAN
1110½ Hay Street, Fayetteville, NC 28305 • preppub@aol.com • (910) 483-6611

OBJECTIVE

To offer excellent technical skills and knowledge to an organization that can use a self-motivated professional with a high level of initiative and drive along with dedication to producing excellent results during critical aircraft quality control and inspection activities.

EDUCATION & TRAINING

Pursuing a Bachelor's degree at East Central University, Ada, OK; 3.6 GPA.
Excelled in U.S. Air Force leadership school as well as in technical schools and extensive on-the-job training for specialists in Non-Destructive Inspection (NDI) aircraft maintenance.

TECHNICAL EXPERTISE

Equipment: Operate and troubleshoot technical equipment which includes: FAS 2C spectrometer; Spectrom, Spectro Jr., and Lorad x-ray units; magnetic particle machines; penetrant lines, USN 50 ultrasound machines; and 19 E II, Hocking, and Nortec machines.
Computers: Operate a wide variety of computer programs used for record keeping and documentation of quality control and supply operations.
Aircraft: Work on B-1B, C-130, A-10, KC-135, C-141 and C-17 aircraft and TF-34 engines.

EXPERIENCE

Have earned a reputation as a skilled professional with an eye for detail and dedication to ensuring safety and quality of multimillion-dollar aircraft, U.S. Air Force:
SUPERVISORY QUALITY CONTROL INSPECTOR. Altus AFB, OK (2005-present). Credited with performing at a level "head and shoulders above" my peers, have saved the Air Force in excess of $3.5 million by applying technical expertise leading to early detection of cracks which can then be repaired and aircraft quickly returned to service.

- Support A-10 and C-130 aircraft while supervising as many as five people.
- Run the Joint Oil Analysis Program (JOAP) lab: processed more than 500 samples with no flight cancellations due to oil-related metal wear and problems.
- Interpret and evaluate indications of defects in aircraft, support equipment, components, and pressurized systems using a wide range of industrial radiography, ultrasonic, eddy current magnetic particle, and fluorescent dye penetrant methods.
- Utilize the Process Control Automated Management System (PCAMS) to document procedures; monitor radiation areas; am a certified Hazardous Material Handler.

NDI APPRENTICE. Pope AFB, NC (2003-05). Evaluated as an "exceptional performer," performed the full range of penetrant, magnetic, ultrasound, eddy current, and x-ray inspections of B-1B and C-130 aircraft and aerospace ground equipment.

- Became familiar with AF technical order procedures and assisted engineers and contractors during the modification of dye penetrant line procedures.
- Quickly earned the respect of my superiors and peers for my outstanding technical knowledge and contributions in support actions including supply and bench stock.
- Developed and implemented procedures which improved supply requisitioning and accounting records maintenance 60%; reduced inventory maintenance 30% by creating a new hazardous material ordering and tracking system for 22 critical items.

SUPPLY TECHNICIAN. Barksdale AFB, LA (2000-03). Awarded an Achievement Medal and evaluated as a diligent and detail-oriented professional who made significant contributions, solved supply problems, monitored repair parts processing documentation and records while inputting changes, deletions, and additions to maintenance records.

- Revitalized an Awaiting Parts (AWP) programs and cut both the number of sidelined aircraft and period of time required to get the parts in half.

PERSONAL

Secret security clearance. Am a hard worker who gets results. Excellent references.

Date

Exact Name of Person
Title or Position
Name of Company
Address
City, State, Zip

SUPERVISORY TECHNICIAN and QUALITY INSPECTOR

Dear Exact Name of Person: (or Dear Sir or Madam if answering a blind ad.)

I would appreciate an opportunity to talk with you soon about how I could contribute to your organization through my experience in quality assurance inspection and management as well as through my technical skills.

While serving my country in the U.S. Navy, I quickly earned a reputation as a proficient technician and natural leader and was selected for advanced training. I received more than 940 hours of technical and management training and became known as a "technical expert" on hydraulic, air conditioning, refrigeration, and high pressure air and oxygen systems.

I have been involved in quality assurance program management since the beginning of my Navy career and have saved thousands of dollars in government funds by conducting QA programs which consistently produced high quality maintenance that resulted in "no reworks." My experience has extended to all aspects of QA program management from supervising personnel, ensuring maintenance and repair, and record-keeping operations.

Through my attention to detail, managerial and supervisory experience, and participation in quality assurance operations, I feel that I offer a "track record" of performance which would make me a valuable asset to your organization.

I hope you will welcome my call soon to arrange a brief meeting at your convenience to discuss your current and future needs and how I might serve them. Thank you in advance for your time.

Sincerely yours,

Donovan Hunter

Alternate last paragraph:
I hope you will call or write soon to suggest a time convenient for us to meet and discuss your current and future needs and how I might serve them. Thank you in advance for your time.

DONOVAN HUNTER

1110½ Hay Street, Fayetteville, NC 28305 • preppub@aol.com • (910) 483-6611

OBJECTIVE

To benefit an organization that can use an experienced specialist who offers skills both in conducting inspections and managing quality assurance programs while applying expertise related to personnel supervision, equipment maintenance, and record keeping.

TRAINING

Excelled in more than 940 hours of technical and management training programs including:
machinist mate "A" school emphasizing shipboard systems — 480 hours
hydraulics/heating/air conditioning/high-pressure air and oxygen systems — 360 hours
Quality Assurance (QA) inspection procedures — 40 hours
calibrating gages and temperature monitoring devices — 40 hours
inspection techniques especially for oxygen systems — 24 hours

EXPERIENCE

Earned advancement based on leadership and knowledge, U.S. Navy, China Lake, CA:
SUPERVISORY TECHNICIAN and **QUALITY INSPECTOR.** (2005-present). Oversee the performance of seven specialists and was known as an "expert technician" on critical systems including hydraulics, air conditioning and refrigeration, and high-pressure air and oxygen systems.

- Polished my inspection skills as a key member of the Quality Assurance Program team.
- Saved the Navy thousands of dollars by assuring work was performed up to standards which resulted in eliminating the need for "reworks" of any jobs I oversaw.

QUALITY ASSURANCE INSPECTOR and **ENGINEERING TECHNICIAN.** (2004-05). Advanced to the QA management level after participating in intensive on-the-job training and gaining increased recognition as a highly skilled technician on hydraulic and high-pressure air systems.

- Was widely known as an "expert" on the Dresser-Rand high-pressure air compressor.
- Refined my management skills while achieving a high standard of production resulting in "zero reworks." Polished my abilities relative to managing time and handling pressure while "juggling" the demands of training and excelling in my regular responsibilities.

DAMAGE CONTROL AND FIREFIGHTING PROGRAMS TECHNICIAN. (2003-04). Was placed in charge of special-use equipment while involved in maintaining hydraulic and high-pressure systems. Rapidly expanded technical knowledge and earned advanced roles.

Received training based on technical and leadership qualities, U.S. Navy, Norfolk, VA:
ENGINEERING TECHNICIAN, AIR CONDITIONING SYSTEMS EXPERT, and **TECHNICAL SUPERVISOR.** (2002-03). Handled a variety of roles while earning a reputation as an outstanding performer, quick learner, and professional offering exceptional attention to detail and management abilities.

- Selected to receive training in air conditioning/refrigeration, was the "subject matter expert" and managed an upgrade project.
- Managed air conditioning system shutdown during the submarines's "decommissioning" and completed work in a timely manner thereby saving thousands of dollars. Coordinated a project to unload and inventory the ship's damage control and firefighting gear.

TECHNICAL EXPERTISE

Through experience and training, am highly proficient in QA inspection; troubleshooting high-pressure air systems including Rand systems, and equipment, and troubleshooting air conditioning/refrigeration units.

PERSONAL

Secret security clearance. Well-developed leadership and motivational skills.

Date

Exact Name of Person
Title or Position
Name of Company
Address
City, State, Zip

TEAM LEADER & TRAINING SPECIALIST

Dear Exact Name of Person: (or Dear Sir or Madam if answering a blind ad.)

I would appreciate an opportunity to talk with you soon about how I could benefit your organization through my outstanding work ethic and persistence in striving for high standards of professionalism and productivity in everything I attempt.

You will see by my enclosed resume that I excel in developing and running training programs which produce qualified and effective employees. I am a quick learner who easily absorbs new methods and procedures and then can take that information and present it to others clearly and concisely. In my present job as a Training Specialist and Team Leader, I have been very effective in helping put together a team of dedicated workers who are achieving high levels of productivity for this manufacturer.

In every position I have held, training and instructing others was at least a part of my responsibilities, and in each case I applied my leadership and knowledge to increase productivity and efficiency. Having worked in a nuclear power plant as a decontamination specialist, built a successful day care center from the ground up, and managed a multifaceted food service program among other roles, I have demonstrated that I am adaptable and versatile.

My strongest abilities are in building teams, providing quality training, and motivating others to excel and maximize their own individual talents. I possess sound judgment and problem-solving skills along with a reputation as an articulate public speaker and training specialist.

I hope you will welcome my call soon to arrange a brief meeting at your convenience to discuss your current and future needs and how I might serve them. Thank you in advance for your time.

Sincerely yours.

Kimberly Curtis

Alternate last paragraph:
I hope you will call or write me soon to suggest a time convenient for us to meet and discuss your current and future needs and how I might serve them. Thank you in advance for your time.

KIMBERLY CURTIS
1110½ Hay Street, Fayetteville, NC 28305 • preppub@aol.com • (910) 483-6611

OBJECTIVE

To contribute to an organization that can benefit from my skills in organizing, supervising, and training others to achieve outstanding results and increased productivity/profitability while assuring the highest standards of safety and quality control.

EXPERIENCE

TEAM LEADER & TRAINING SPECIALIST. Lincoln Corporation, Shepherdstown, WV (2005-present). Selected to train and develop a team of skilled employees from the start-up of a new manufacturer, maintain certification in the operation of nine different pieces of production machinery and ensure production goals are met.

- Screened and selected new employees; personally trained more than 50 people.
- Increased productivity 40% through numerous suggestions which were accepted.
- Oversaw the Continuous Improvement (CI) program which targets methods of increasing productivity and presented weekly CI training classes.
- Counseled personnel on the ways to be effective and productive team members and how to provide technical support services to other departments.
- Excelled in communicating with other team leaders, shift supervisors, and staff members so that my team always had the supplies and equipment to do the job.
- Completed detailed technical work including entering calibration data into a computer database, reviewing data on Excel spreadsheets, and assigning ID numbers to gauges.

NUCLEAR DECONTAMINATION SPECIALIST. West Virginia Nuclear Power Plant, Shepherdstown, WV (2002-05). Handled a wide range of activities including maintaining safety standards, cleaning controlled spills, disposing of radioactive waste, operating forklifts, and preparing parts and equipment for use by pipe fitters.

- Conducted informative training sessions as a member of the Safety Committee and ensured employees were aware of applicable NRC (Nuclear Regulatory Commission) and OSHA (Occupational Safety and Health Agency) regulations.
- Taught myself to be constantly aware and communicate with others on ways to work smartly while limiting exposure to radioactive materials.

ASSISTANT MANAGER. AAFES, Italy (2000-01). Supervised over 100 employees in a food service program which supported three food court restaurants and two schools while taking care of functions including coordinating vendor activities, processing payroll, scheduling, and maintaining employee personnel records.

- Developed a cleaning schedule accepted for use in all five facilities and which resulted in a 100% pass rate for all health and safety inspections.
- Oversaw a project to build a pizza restaurant from the construction phase, to screening and hiring employees, to running an orientation program which earned the praise of the District Manager as the best and most thorough he had seen.

EDUCATION & TRAINING

Completed extensive college and corporate training in areas including:

decontamination | team dynamics | hazardous materials handling
radwaste evolutions | lift truck operation | radioactive material shipment
nuclear fuel programs | confined space | Total Quality Management (TQM)
respiratory protection | firewatch duties | competitive manufacturing
chemical control | spill control | environmental health safety
decontamination/frisking for deconners | team effectiveness

PERSONAL

Quickly absorb and apply new information and procedures. Skilled public speaker and instructor. Have a reputation for persistence. Offer outstanding references.

Power generation equipment

Date

Exact Name of Person
Title or Position
Name of Company
Address
City, State, Zip

TECHNICAL INSPECTOR

Dear Exact Name of Person: (or Dear Sir or Madam if answering a blind ad.)

I would appreciate an opportunity to talk with you soon about how I could contribute to your organization through my outstanding technical and supervisory skills as well as through my specialized experience conducting inspections and overseeing quality control activities.

I was handpicked for my present job as a Power Generation Technical Inspector at Fort Campbell, KY. In this role I have become widely recognized as the best and most knowledgeable inspector for an organization which supports more than 35 units of varying sizes. During my more than 13 years of military experience, I have become familiar with meeting short deadlines and heavy work loads in ways which increase productivity and decrease equipment downtime.

As you will see from my enclosed resume, I offer a reputation as a skilled maintenance technician with strong abilities in troubleshooting and equipment repair. I am confident that I offer a blend of communication, motivational, and leadership abilities as well as technical and mechanical skills which combine to make me a very adaptable professional who easily masters new technical information.

I hope you will welcome my call soon to arrange a brief meeting to discuss your current and future needs and how I might serve them. Thank you in advance for your time.

Sincerely,

Otto John

Alternate last paragraph:
I hope you will call or write me soon to suggest a time convenient for us to meet and discuss your current and future needs and how I might serve them. Thank you in advance for your time.

OTTO JOHN

1110½ Hay Street, Fayetteville, NC 28305 • preppub@aol.com • (910) 483-6611

OBJECTIVE

To contribute my outstanding abilities related to troubleshooting, maintaining, overhauling, and testing power generation equipment to an organization that can use my specialized experience in inspection and quality control as well as leadership, supervision, and resource management.

AREAS of EXPERTISE & CLEARANCE

Operate analog and digital diagnostic equipment while completing technical troubleshooting.
Apply troubleshooting to wiring diagrams and schematics on Generator Power Units.
Use equipment including phase rotation meters, energy analyzers, clamp-on amp meters, voltage continuity testers, portable load banks, maintenance tools, ohmmeters, voltmeters, analog and digital circuit testers, multimeters, and voltage probes.
Operate equipment including up to 10K forklifts, D5 bulldozers, and Scamp cranes.
Troubleshoot Scamp cranes and arc welding machines.
Maintain diesel engines to include adjusting valves, carburetors, ignition points, voltage regulators, control circuits, and solenoids.
Was entrusted with a **Secret** security clearance.

EDUCATION & TRAINING

Studied Criminal Justice, Hopkinsville Community College, KY, 1998-01.
Completed training programs which included:
Environmental Certification – handling hazardous material/proper disposal of hazardous waste
Equal Opportunity Course – identifying and responding to discrimination or harassment
Ground Power Unit Training – maintenance on turbine engine systems

EXPERIENCE

TECHNICAL INSPECTOR. U.S. Army, Fort Campbell, KY (2005-present). Have become widely recognized as the best and most knowledgeable power generation inspector within the 101st Airborne Division which supports 35 units of varying size; supervise 11 technicians.

- Completed initial and final quality control inspections on power generation equipment once it had been repaired.
- Troubleshot, isolated, and identified the causes of equipment malfunctions.
- Served as the administrator for quality assurance functions with responsibility for equipment valued in excess of $2.5 million.Achieved an impressive 95% readiness rate for the equipment during its preparation for real-world combat missions.
- Was handpicked from a pool of seven qualified and experienced personnel for this role.
- Praised for my ability to motivate and train others, demonstrated a talent for developing the strengths of subordinate personnel and for developing effective cross-training programs. Was a key player in the unit's recognition with a commendable rating on a state-level environmental inspection.

POWER GENERATOR MECHANIC. U.S. Army, Fort Campbell, KY (2004-05). Built a reputation as a thoroughly knowledgeable technician and supervisor while maintaining and overhauling power generation equipment.

- Became skilled in repairing small motor generators; maintained and repaired diesel engines. Maintained records as required by applicable technical requirements.
- Assisted in overhauling engines and generators, replacing worn and defective parts, and making final adjustments prior to reassembly.
- Was awarded an Army Commendation Medal for my support during the War on Terror.

PERSONAL

Highly skilled troubleshooter. Am very effective at motivating others to accomplish assigned tasks and to build their skills and strengths.

Waste water management

Date

Exact Name of Person
Title or Position
Name of Company
Address
City, State, Zip

TECHNICAL INSTRUCTOR

Dear Exact Name of Person: (or Dear Sir or Madam if answering a blind ad.)

I would appreciate an opportunity to talk with you soon about how I could benefit the Northwestern Water Works through my education, knowledge, and supervisory skills. I am especially interested in receiving your consideration for the positions of Laboratory Supervisor and Chemist.

You will see by my enclosed resume that I have a B.S. degree in Biology and a B.S. in Business Administration along with additional graduate-level course work at Casper College. This program is in the area of medical education development and allows students an opportunity to complete a concentrated course showing them the academic, physical, and mental stresses of medical school. I earned certification for completing this program two different summers.

My versatile background includes acting as a coordinator for Western Wyoming Community College where I helped administer a program for NWW employees from a three-state area in the pipe and water meter distribution courses. This program, which I have been involved in since 2003 on a part-time basis, instructs students on the American Waste Water Association (AWWA) standards and procedures. I am also a Technical Instructor for the Industrial Maintenance program at WWCC which leads to certification in areas including lift truck operation, safety in the work place, math and measurements, and blueprint reading.

In addition to the experience and education mentioned above, I also offer well-developed supervisory skills refined as a military officer. After several years in the field of transportation operations, I requested and was accepted for the field of supply management. My military experience allowed me opportunities to build a reputation as a talented developer of comprehensive training plans as well as a successful manager of human, material, and fiscal resources.

I offer a well-rounded background directly related to NWW operations along with the ability to quickly master new ideas and procedures. I feel that through my versatility and ability to adapt to new things I could quickly become a valuable asset to your organization.

I hope you will welcome my call soon to arrange a brief meeting at your convenience to discuss your current and future needs and how I might serve them. Thank you in advance for your time.

Sincerely yours,

Henry Paul

HENRY PAUL

1110½ Hay Street, Fayetteville, NC 28305 • preppub@aol.com • (910) 483-6611

OBJECTIVE

To offer a versatile background which includes experience in the areas of developing training programs and instructing technical subjects as well as managing supply and transportation operations while becoming known for my versatility and intellectual abilities.

EDUCATION

B.S., Biology, Western Wyoming Community College, Rock Spring, WY.
B.S., Business Administration, Casper College, Casper, WY.
Completed graduate-level medical education development programs, Casper College, 2005 and 2000; emphasis was on gross anatomy, histology, biochemistry, physiology, and dental lab operations while ensuring participants were subjected to the mental, physical, and intellectual stresses of medical school.

TRAINING

Completed extensive training programs for military executives with an emphasis on the refinement of managerial and supervisory skills as well as specific programs in the areas of transportation, supply, and civil affairs management.

EXPERIENCE

TECHNICAL INSTRUCTOR, QUALITY ASSURANCE DEPARTMENT. Western Wyoming Community College, Rock Springs, WY (2004-present). Am applying my communication and technical skills as a coordinator for the American Waste Water Association (AWWA) pipe and water meter distribution courses which are given to NWW employees from a three-state area.

- Provided instruction in the Industrial Maintenance program which included lift truck operation, safety, math and measurements, and blueprint reading.
- Was selected to teach courses jointly sponsored by WWCC and the City of Rock Springs: taught construction, basic math/measurements, and blueprint reading courses in the construction class.

QUALITY CONTROL SUPERVISOR. NWW, Inc., Rock Springs, WY (2003). Ensured the quality of work performed in projects at Yellowstone National Park: saw that landscaping projects were properly completed including compliance work reference, material quality/ quality, and safety data sheets as well as calculating costs.

FULL-TIME STUDENT. Casper College, Casper, WY (1997-03).

SUPPLY OPERATIONS MANAGER. National Guard, Jackson, WY (1989-96). Was selected for attendance at numerous management training programs in the transportation and supply fields while also completing college course work in pre-dentistry and pre-pharmacy.

Highlights of other experience: Applied a variety of skills and abilities while handling such full- and part-time jobs as maintaining rental units for a family property management company, volunteering at the Memorial Hospital dental clinic, and totally rebuilding a mobile home which had been destroyed by fire.
As a U.S. Army Officer, handled the training and administrative support for a 25-person supply center supporting a National Guard unit.

- Developed and implemented training programs in areas including maintenance, personnel, logistics, and workplace safety.
- Planned and prepared all aspects of training as well as overseeing security for a large-scale supply support organization.

PERSONAL

Am a versatile and adaptable professional who can quickly learn new procedures.

Automated information systems

Date

Exact Name of Person
Title or Position
Name of Company
Address
City, State, Zip

TOTAL QUALITY MANAGEMENT (TQM) COORDINATOR

Dear Exact Name of Person: (or Dear Sir or Madam if answering a blind ad.)

I would appreciate an opportunity to talk with you soon about how I could contribute to your organization through my management and supervisory skills as well as through my technical abilities in the area of automated information systems, training development, assets accountability, and inventory control.

As you will see from my resume, I have earned a reputation as one of the U.S. Navy's experts in implementing Total Quality Management (TQM) to reduce costs and improve efficiency. During my management experience as an officer, I have come to believe strongly that continuous training of human resources is often the key to an organization's superiority in the marketplace. Skilled in diagnosing training needs and conceptualizing methods needed for improvement, I am respected among employees as a manager who is skilled at balancing company goals and employee needs.

You would find me in person to be a highly motivated young executive who takes pride in my ability to communicate visions and motivate personnel to achieve them. I can provide outstanding personal and professional references. I am cheerfully willing to relocate.

I hope you will welcome my call soon to arrange a brief meeting at your convenience to discuss your current and future needs and how I might serve them. Thank you in advance for your consideration.

Sincerely yours,

Marcus Jones

Alternate last paragraph:
I hope you will call or write me soon to suggest a time convenient for us to meet and discuss your current and future needs and how I might best serve them. Thank you in advance for your consideration.

MARCUS JONES

1110½ Hay Street, Fayetteville, NC 28305 • preppub@aol.com • (910) 483-6611

OBJECTIVE

I want to contribute to an organization that can use a safety and quality assurance expert.

EXPERIENCE

U.S. NAVY, Commissioned Officer (1999-present):

TOTAL QUALITY MANAGEMENT (TQM) COORDINATOR. U.S. Naval Reserve Readiness Center, Corpus Christi, TX (2005-present). With a reputation as a gifted communicator with a knack for solving stubborn problems, have become the "resident expert" on statistical process control while earning widespread praise for my innovative approach to teaching TQM and reducing employee fear of statistical analysis.

- ***Total Quality Management***: Developed a TQM approach that has become the model for all naval reserve centers in TX, LA, FL, and GA; drafted the five-year TQM plan for this 800-person organization; have become respected as one of the Navy's experts in using TQM to reduce costs and improve efficiency.
- ***Automated Information Systems***: Undertook a comprehensive analysis of computer software/hardware and reorganized the AIS program to improve data security, increase efficiency, and better utilize training dollars.

TRAINING OFFICER. Naval Reserve Readiness Center, Corpus Christi, TX (2004-05). Was specially selected for this "plum" assignment at the largest naval center in the northeast; administered a $4.2 million budget, managed customer service, and supervised a commercial teleticketing operation while overseeing training/travel of 1,700 personnel.

- ***Spreadsheet development***: Developed a computerized spreadsheet for tracking travel expenditures and implemented an electronic information exchange system which was later adopted as "the system" for 17 other centers.
- ***Management***: Took over a center ranked #10 in the northeast and, through my communication and problem solving skills, led the center to be ranked #2.

NAVIGATOR. U.S.S. North Carolina, homeport in Wilmington, NC (2003-04). Was the first division-level officer on this ship ever promoted to this job, usually reserved for much more senior personnel with proven judgment and decision-making ability; reported directly to the ship's commanding officer.

- ***Planning***: In this no-room-for-error job, planned the safe navigation of a 16,000-ton ship through the world's most challenging waters.
- ***Project management***: Implemented a new satellite-based system for navigation.

ADMINISTRATOR. U.S.S. Constellation, homeport in San Diego, CA (2002-03). Reorganized the ship's automated data processing procedures and developed advanced training of personnel on new software; decreased paperwork processing time 150%.

COMMUNICATIONS OFFICER. U.S.S. Constellation, homeport in San Diego, CA (1999--02). Responsible for all visual and electronic communications, including security of Top Secret coding material.

- ***Crisis management***: While in the Persian Gulf, took over electronic communications of a 20-ship battle group for 48 hours during a satellite failure.

EDUCATION

Earned **B.S. degree in Finance**, Arizona State University West, Phoenix, AZ, 1999.

- First runner-up in a university-wide political science essay contest.

As a naval officer, excelled in more than two years of **graduate-level** courses including rigorous technical schools. Subjects included shipboard engineering, management and supervision, Total Quality Management (TQM), and electronic-communications, 1999-04.

Organizational effectiveness

Date

Exact Name of Person
Title or Position
Name of Company
Address
City, State, Zip

TOTAL QUALITY MANAGEMENT (TQM) INSTRUCTOR

Dear Exact Name of Person: (or Dear Sir or Madam if answering a blind ad.)

Can you use a talented and energetic young professional who offers a strong interest in pursuing a career which will require creativity and a sense of style while allowing me to utilize my considerable talents in the art community?

You will see from my enclosed resume that I offer a proven ability to handle heavy work loads and "get things done" no matter what the obstacles. I am an expert at managing my time for maximum effectiveness. While excelling in my full-time management position, I earned an associate's degree and a bachelor's degree at night from respected institutions while also finding the time to become a volunteer leader in the arts community in my spare time. I feel certain my management experience, organizational ability, and communications skills are transferable to any industry. I have traveled extensively in Africa and Asia and am accustomed to dealing with people of diverse cultures.

In my most recent job I wore "two hats"; I managed eight technical specialists while controlling a $5 million dollar account, and I was an innovative instructor of Total Quality Management (TQM) at one of the military's busiest airlift centers. While teaching TQM classes to groups of up to 45 individuals, I refined my ability to "sell" ideas while helping people change preconceived and often-hostile impressions of TQM into positive attitudes. I was commended frequently for my salesmanship and for my excellent performance as a "change agent" in teaching people how to apply TQM in every area of the work environment.

I have a strong desire to utilize my versatile professional skills in a job where I can pursue my interest in the arts. I have had the opportunity to coordinate several local fashion shows and am accustomed to speaking in public. As a member of the MS Museum of Art and a volunteer Docent at the Gulfport-Biloxi Museum of Art for the past three years, I have coordinated art exhibits while acquiring expertise in handling, storing, and displaying art objects as well as maintaining art records. As a Docent, I conducted tours and handled extensive public speaking about artists and their works. I have a deep appreciation and passion for art, and I am very effective in passing that love on to others.

I hope you will welcome my call soon to arrange a brief meeting at your convenience to discuss your current and future needs and how I might serve them. Thank you in advance for your time.

Sincerely yours,

Martina Elliott

MARTINA ELLIOTT

1110½ Hay Street, Fayetteville, NC 28305 • preppub@aol.com • (910) 483-6611

OBJECTIVE

To offer my creativity, problem-solving ability, and communication skills to an organization that can use a poised young professional who offers experience in public speaking and well-developed interpersonal skills as well as a reputation for initiative and a drive to excel.

EDUCATION & TRAINING

B.S., Psychology, William Carey College, Gulfport, MS, and **A.S., Logistics**, Community College of the Air Force, 2005; completed both degrees while working full time.
Excelled in training programs including a leadership/management course and a program leading to certification as an **Instructor of Total Quality Management (TQM)** principles.

EXPERIENCE

Developed a reputation for outstanding planning, organizational, and problem-solving skills as well as a talent for motivating and communicating with others effectively, U.S. Air Force:
TOTAL QUALITY MANAGEMENT (TQM) INSTRUCTOR. Keesler AFB, MS (2003-present). Was promoted ahead of my peers to manage a $5 million account and supervise eight technical specialists while ordering, controlling, and storing an $8 million inventory of supplies and equipment used to provide aircraft services.

- In my role as a TQM Instructor for classes of up to 45 people at the world's largest U.S. military base, have excelled as a "change agent" while teaching professionals how to apply TQM concepts in order to improve every aspect of their work environment.
- Controlled operations in 11 flight-line supply points with a total of 256 line items.
- Achieved a 95% inventory accuracy rate while controlling costly spare parts.
- Am respected for my ability to work under pressure while handling multiple projects with constantly changing priorities.

RESOURCES MANAGER. Keesler AFB, MS (2002-03). Supervised two supply specialists; controlled an $8 million account used to acquire essential parts for aircraft and aerospace ground equipment; worked closely with my counterparts at other air force bases to locate needed parts.

- Brought about a 15% reduction in the number of important systems that were out of commission while waiting for needed parts to arrive so repairs could be completed.
- Processed approximately 250 shipments where the equipment was needed to return aircraft grounded at other locations and waiting for critical parts.

SUPPLY COORDINATOR. Keesler AFB, MS (2002). Ensured error-free processing of more than 100 priority requests daily as the main point-of-contact with additional responsibility for quality control and researching supply management listings.

- Achieved a 97% on-time delivery rate while researching and processing approximately 30 part-number issue requests each day and providing quality control support.
- Played a key role in the accurate processing of 500 priority issue requests.
- Received the Southwest Asia Service Medal for my expertise during the War on Terror.

MAINTENANCE/SUPPLY LIAISON. Germany (2000-01). As point-of-contact and advisor, ensured the smooth coordination of activities between maintenance and supply.

- Achieved an 86% equipment availability rate (well above expected standards) and reduced order processing times from 17 to 14 days.

PERSONAL

Founded a group called "The Sibling Group" which meets monthly and allows an opportunity for women to discuss issues related to personal growth and change. Was entrusted with a Secret security clearance. Have a working knowledge of German. Am single and will travel and/or relocate as needed by my employer.

Banking

Date

Exact Name of Person
Title or Position
Name of Company
Address
City, State, Zip

VICE PRESIDENT, COMPLIANCE

Dear Exact Name of Person: (or Dear Sir or Madam if answering a blind ad.)

I would appreciate an opportunity to talk with you soon about how I could contribute to your organization through my extensive management experience in most functional areas of accounting, human resources, operations, and banking.

As you will see from my resume, I have enjoyed a track record of promotion with National Consumer Bank, one of the leading financial institutions in the South. I have often worn multiple "hats" and am known for my ability to oversee complex responsibilities in numerous areas simultaneously. For example, in my current position as a vice president, I oversee the Operations and Compliance areas for the bank, and I actually developed the bank's Deposit Compliance Program.

If you feel your management team could benefit from my in-depth experience, creative problem-solving style, and reputation as a strategist and visionary, I would be delighted to make myself available at your convenience to discuss your needs and goals and how I might help you achieve them. I do wish to point out that I will be relocating to the Houston area in order to be closer to my family who all live in or near Houston.

I hope you will welcome my call soon to arrange a brief meeting to discuss your current and future needs and how I might serve them. Thank you in advance for your time.

Sincerely,

Anne Wade

ANNE WADE

1110½ Hay Street, Fayetteville, NC 28305 • preppub@aol.com • (910) 483-6611

OBJECTIVE

To contribute to an organization that can use an innovative manager who believes that "the sky is the limit" when persistence, creativity, and attention to detail are combined with superior planning, time management, communication, and problem-solving skills.

EXPERIENCE

For more than 14 years, have built a "track record" of accomplishment in positions of increasing responsibility at National Consumer Bank, Virginia Beach, VA:

VICE PRESIDENT, COMPLIANCE. (2000-present). Was promoted to handle additional responsibilities related to consumer compliance while continuing to handle responsibilities described in the Assistant Vice President job below.

- Developed Deposit Compliance Program for consumer law and regulation.
- Personally conduct compliance testing (audits) and the training program; have trained approximately 40 employees in this specific area.
- Have become very experienced in internal auditing for compliance.

ASSISTANT VICE PRESIDENT OF OPERATIONS & HUMAN RESOURCES. (1995-2000). In this highly visible, fast-paced position reporting to the bank president, wore "three hats," balancing multiple responsibilities in human resources, operations, and investments.

- ***Management:*** Directly supervised five people, ensuring that assigned responsibilities were executed in a systematic and effective manner.
- ***Operations:*** In coordination with top management, developed/implemented plans and policies that affected accounting, bookkeeping, and data processing of main office and two branches.
- ***Human Resources:*** Applied my expert knowledge to develop, maintain, and administer all personnel policies as they applied to 40 bank employees; oversaw EEO compliance, recruitment, safety & health.
- ***Benefits Administration***: Oversaw all salary and benefit functions, including 401(k) pension plan and Blue Cross/Blue Shield health plan.
- ***Finances***: Managed an investment portfolio utilizing excess funds per day while efficiently planning and administering the department's budget.
- ***Training:*** Coordinated in-house programs on personnel policies; organized training on compliance with demand deposit, direct deposit, bank privacy, and other regulations.

OPERATIONS OFFICER. (1991-1995). Excelled in directing all aspects of the Operations Department because of my versatile management skills; supervised five employees.

- Reviewed surveys of community banks and made recommendations on competitively pricing various banking products.
- As a member, Strategic Planning Committee, offered input on personnel and operations.
- Managed all day-to-day bookkeeping functions and monitored the bank's cash position, making investments or borrowing funds as appropriate.

HEAD BOOKKEEPER. (1989-1991). Ensured the highest standards of customer service while supervising and reviewing the work of five assistants; balanced general ledger accounts; processed overdrafts, returns items, and ACH debits and credits.

EDUCATION

B.S., Accounting, Virginia State University, Virginia Beach, VA, 1991.
Excelled in courses in Accounting, Introduction to Computers, Banking & Finance, Business Communication, Supervision, and Principles of Management.
Attended seminars on sexual harassment, interviewing & hiring, state and federal wages, personnel policies (developing/implementing), public speaking, and check processing.

PERSONAL

Self-motivated, dedicated professional with a reputation as a team leader.

Purification of saltwater

Date

Exact Name of Person
Title or Position
Name of Company
Address
City, State, Zip

WATER PURIFICATION ENGINEER

Dear Exact Name of Person: (or Dear Sir or Madam if answering a blind ad.)

With the enclosed resume, I would like to make you aware of my strong technical skills and natural leadership ability as well as the background of excellence in Water Treatment and Purification Systems operation, management, and training which I could put to work for your organization.

As you will see from my resume, I am currently excelling as a Water Treatment Engineer for the U.S. Army, where I train and direct the work of eight employees in the development of water sources and analysis of raw and treated water, as well as the operation and maintenance of water treatment equipment. In addition to overseeing the security, maintenance, and accountability of $750,000 worth of equipment, I cross-trained petroleum and ammunition supply employees to perform water treatment functions, increasing the versatility and effectiveness of these personnel. While stationed in Afghanistan, I was the first water treatment specialist to purify saltwater using the Reverse Osmosis Water Purification Unit (ROWPU), and I trained ten Afghani personnel to operate various types of water treatment equipment.

A graduate of the Level I, II, III, & IV Wastewater Treatment Plant Operator course at University of Kentucky-Hopkinsville Community College, I have also completed numerous military technical and leadership training courses. These included the U.S. Army Quartermaster School, Water Treatment Specialist Basic Non-Commissioned Officers Course (BNCOC), Primary Leadership Development Course (PLDC), Jumpmaster Course, and Master Fitness Trainer Course.

Throughout my military career, I have demonstrated strong leadership and training skills, as well as the ability to quickly master new and complex technical information. My energy, drive, and enthusiasm have allowed me to motivate personnel under my supervision to achieve excellence both personally and professionally, and I have built a reputation as an articulate leader with unlimited potential for advancement.

If you can use a highly skilled professional whose leadership and technical abilities have been proven in challenging environments worldwide, I hope you will welcome my call soon when I try to arrange a brief meeting to discuss your goals and how my background might serve your needs. I can provide outstanding references at the appropriate time.

Sincerely,

Terrence Karl

TERRENCE KARL

1110½ Hay Street, Fayetteville, NC 28305 • preppub@aol.com • (910) 483-6611

OBJECTIVE

To benefit an organization that can use an articulate young professional with exceptional technical, organizational, and leadership abilities who offers a background in water treatment operations and management, supervision and training of personnel, and fitness training.

EDUCATION

Completed the **Wastewater Treatment Plant Operator Level I, II, III, & IV** courses, University of Kentucky-Hopkinsville Community College, KY, 2004.
Excelled in military leadership and technical skills training courses, including the Primary Leadership Development Course (PLDC), U.S. Army Quartermaster School, Water Treatment Specialist Course, Basic Non-Commissioned Officers Course (BNCOC), Army Institute for Personal Development, Jumpmaster Course, Pathfinder Course, Air Movement Operations Course (for Air Transport of Hazardous Materials), and Master Fitness Trainer Course.

TECHNICAL SKILLS

Water Treatment: Skilled in the operation of ultraviolet filtration devices, osmosis (Erdlator) & reverse osmosis (ROWPU) purification units.
Materials handling and other equipment: Qualified to operate 4, 6, and 10K forklifts, 5-ton cranes, 40-foot trailers, excavation vehicles, and Global Positioning Systems (GPS).
Computers: Familiar with many popular computer operating systems and software, including Microsoft Word, Excel, and PowerPoint; and others.

LICENSES

Preparing to test for the Kentucky Wastewater Treatment Plant Operator's License.

EXPERIENCE

Was selected for advanced technical training and promoted to positions of increasing responsibility while serving in the U.S. Army, 1997-present:

2001-present: WATER PURIFICATION ENGINEER. Fort Campbell, KY and Afghanistan. Supervise and train as many as eight personnel in development of water sources, analysis of raw and treated water, and maintenance of water treatment equipment.

- Oversee the security, maintenance, and accountability of equipment valued at more than $750,000; cross trained in aircraft refueling, including HAZMAT and safety issues.
- Known as a technical expert on all issues related to water purification and distribution.
- While stationed in Afghanistan, was the first water treatment specialist to utilize the Reverse Osmosis Water Purification Unit (ROWPU) to purify sea water; trained ten Afghani personnel in the operation of various types of water purification equipment.
- Cited in official performance appraisals as "instrumental in setting up water point supply and distribution" and "[ensuring] that the water team monitored and enforced quality assurance of all water distributed" during a major exercise.
- A natural leader, am sought out by managers, peers, and those under my supervision for expert leadership, technical, and tactical advice; demonstrated high levels of compassion for the morale and welfare of my employees, both on and off duty.

1997-01: LOGISTICS SUPERVISOR. Fort Polk, LA. Provided training in ammunition storage, accountability, and safety to employees under my supervision, as well overseeing the maintenance of ammunition, vehicles, and equipment assigned to the unit.

- Was instrumental in directing and participating in vehicle and equipment maintenance that resulted in achieving a perfect score of 100% during a major inspection.

PERSONAL

Received a number of prestigious awards for my exemplary performance, including an Army Commendation Medal, two Army Achievement Medals, and the Good Conduct Medal, as well as a Humanitarian Award for providing assistance to victims of Hurricane Phillip.

Payroll and tax auditing

Date

Exact Name of Person
Title or Position
Name of Company
Address
City, State, Zip

WORKER'S COMPENSATION AUDITOR

Dear Exact Name of Person: (or Dear Sir or Madam if answering a blind ad.)

With the enclosed resume, I am writing to express my interest in exploring employment opportunities with your organization and make you aware of my background related to tax auditing, fraud investigation, and claims investigation.

As you will see from my resume, I am working as a Worker's Compensation Auditor as an independent subcontractor, and I have worked for insurers. As a skilled payroll and tax auditor for compliance and assessment purposes, I am applying my expert knowledge of Wyoming laws as well as my knowledge of general liability, garage liability, and Worker's Comp.

In my previous experience with the Employment Security Commission of Wyoming, I excelled as a Fraud Investigator, Tax Auditor, and Chief Claims Investigator. When I took over the Cheyenne office of ESC as Chief Claims Investigator, I inherited an inefficient operation. On my own initiative, I studied systems used in CO, MT, and NE, and I reengineered internal systems so that they became models of efficiency. While managing 18 investigators and 26 others, I made numerous changes which resulted in the investigators' caseload increasing from 15 monthly to 150 monthly while increasing detection of fraud from $235,000 a year in fraud overpayments to $6 million detected yearly.

I am skilled in working with the general public as well as organizations including the SBI, IRS, and numerous state agencies.

If you can make use of my considerable experience and skills, please contact me to suggest a time when we might meet to discuss your needs. I am available to travel and/or relocate according to your needs, and I can provide outstanding personal and professional references at the appropriate time.

Sincerely,

Terence Walker

TERENCE WALKER

1110½ Hay Street, Fayetteville, NC 28305 • preppub@aol.com • (910) 483-6611

OBJECTIVE

To contribute to an organization that can use an experienced professional who offers extensive experience in tax auditing, fraud investigation, and claims investigation along with expertise related to Worker's Compensation and expert knowledge of WY laws.

EDUCATION

Bachelor of Arts (B.A.), Economics with minor in Psychology, University of Wyoming, Laramie, WY, 1978.

Extensive training in fraud investigations and auditing techniques received at the WY Justice Academy, Laramie County Community College, Cheyenne, WY.

EXPERIENCE

WORKER'S COMPENSATION AUDITOR. (2005-present). As an independent subcontractor, am involved in auditing business, tax, and accounting records for worker's compensation and employer general liability policies to assure compliance with WY laws and insurance industry standards.

- Based on audits, assign/reassign classification codes for compliance/rating purposes.
- Insurers for whom I have done work include Auto Owners, The Workers Corp., General Liability, The Wyoming Group, Yellowstone's Insurance Company, and others.
- As a skilled payroll and tax auditor for compliance and assessment purposes, apply expert knowledge of general liability, garage liability, Worker's Comp.

Previously excelled in a track record of promotion, Employment Security Commission of WY:

1991-04: TAX AUDITOR. Green River, WY. After requesting reassignment to Green River for family reasons, audited business accounting and tax records for payroll reporting compliance. Interpreted and made independent application of law, rules, and regulations relative to audit data for high- and low-volume payrolls in accordance with auditing standards.

- Audited businesses with multiple state operations; handled tax liability investigations, collections, judgments, delinquent tax reports, bankruptcy, estate cases, asset seizures, and auctions. Served tax liens, negotiated payment plans and settlements, and served as state agent for tax criminal and civil matters in a court of law.
- Specialized in detection of hidden wages and tax liability investigations; served judgements, made seizures, served tax assessments; testified as witness in court.

1987-91: CHIEF CLAIMS INVESTIGATOR. Cheyenne, WY. Took over management of an inefficient and archaic operation, and transformed it into a respected and highly productive organization; managed 18 investigators, one administrative supervisor, one secretary, two supervisors, and 26 others involved in receivables and payables activities.

- Upon taking over operation, studied similar organizations in CO, MT, and NE; then reengineered internal systems to eliminate duplicate tasks.
- Moved the operation from $235,000 a year in fraud overpayments discovered to the detection of $6 million yearly.
- Implemented a computer cross match program utilizing probability techniques which was sent to employers and which emphasizes self-reporting; this boosted taxpayer satisfaction with the system while freeing up investigators' time.
- Through efficiencies which I created and implemented, the caseload of investigators went from 15 per month to 150 per month; provided more support for investigators.

Other experience: FRAUD INVESTIGATOR. ESC of WY. Casper, WY (1983-87). Determined eligibility for unemployment insurance benefits and investigated suspected fraud.

PERSONAL

Served as President of homeowners association for five years. Enjoy coaching youth soccer.

PART THREE:
Applying for Federal Government Jobs

You may already realize that applying for a federal government position requires some patience and persistence in order to complete rather tedious forms and get them in on time. Depending on what type of federal job you are seeking, you may need to prepare an application such as the SF 171 or OF 612, or you may need to use a Federal Resume, sometimes called a "Resumix," to apply for a federal job. But that may not be the only paperwork you need.

Many Position Vacancy Announcements or job bulletins for a specific job also tell you that, in order to be considered for the job you want, you must also demonstrate certain knowledge, skills, or abilities. In other words, you need to also submit written narrative statements which microscopically focus on your particular knowledge, skill, or ability in a certain area. The next few pages are filled with examples of excellent KSAs. If you wish to see many other examples of KSAs, you may look for another book published by PREP: "Real KSAs--Knowledge, Skills & Abilities--for Government Jobs."

Although you will be able to use the Federal Resume you prepare in order to apply for all sorts of jobs in the federal government, the KSAs you write are particular to a specific job and you may be able to use the KSAs you write only one time. If you get into the Civil Service system, however, you will discover that many KSAs tend to appear on lots of different job announcement bulletins. For example, "Ability to communicate orally and in writing" is a frequently requested KSA. This means that you would be able to use and re-use this KSA for any job bulletin which requests you to give evidence of your ability in this area.

What does "Screen Out" mean? If you see that a KSA is requested and the words "Screen out" are mentioned beside the KSA, this means that this KSA is of vital importance in "getting you in the door." If the individuals who review your application feel that your screen-out KSA does not establish your strengths in this area, you will not be considered as a candidate for the job. You need to make sure that any screen-out KSA is especially well-written and comprehensive.

How long can a KSA be? A job vacancy announcement bulletin may specify a length for the KSAs it requests. Sometimes KSAs can be 1-2 pages long each, but sometimes you are asked to submit several KSAs within a maximum of two pages. Remember that the purpose of a KSA is to microscopically examine your level of competence in a specific area, so you need to be extremely detailed and comprehensive. Give examples and details wherever possible. For example, your written communication skills might appear more credible if you provide the details of the kinds of reports and paperwork you prepared.

KSAs are extremely important in "getting you in the door" for a federal government job. If you are working under a tight deadline in preparing your paperwork for a federal government position, don't spend all your time preparing the Federal Resume if you also have KSAs to do. Create "blockbuster" KSAs as well!

FEDERAL RESUME OR RESUMIX

CIVIL ENGINEERING SAFETY MANAGER

SEAN O'CONNER
SSN: 012-34-5678

1110 ½ Hay Street
Fayetteville, NC 28305
(910) 483-6611
Vacancy Announcement Number:

Country of Citizenship: U.S.A.
Veterans' Preference: Veterans Readjustment Appointment
Reinstatement Eligibility: N/A
Highest Federal Civilian Grade Held:

SUMMARY

Offer strong leadership and technical skills, well-developed planning and organizational abilities, and a track record of accomplishment in programming and design of construction projects, supervision and training, and project management.

EXPERIENCE

ENGINEERING SAFETY MANAGER. Air National Guard, Fritter Air Field, Tulsa, OK (2000-present).
Supervisor: Lt. Colonel Francis Sweeney (111) 222-3333
Pay grade: Major (O-4) **Hours worked per week:** 40+
Duties: Supervise and train up to four managers and as many as 104 personnel. Oversee all aspects of facility maintenance and repair, while also handling the duties of the Deputy Civil Engineer, directly handling programming and design of essential facility construction projects, including a critical design for a new $14 million hangar. As Senior Engineer, hold final accountability for a 25-facility physical plant covering more than 350,000 square feet with a value in excess of $125 million. Direct planning and programming of construction programs totaling $50 million, while single-handedly managing a design and construction program comprising 12 major projects totaling more than $25 million. Exceptional staff development and training resulted in 20 personnel under my supervision receiving promotions.

- Due to my specific leadership and execution as Civil Engineering Commander during the Wing's Operational Readiness Inspection, we achieved an overall rating of excellent.
- Negotiated for and obtained a third of the Air National Guard's entire $15 million Real Property Maintenance (RPM) budget in order to provide funding for essential repairs at Fritter Field.

CIVIL ENGINEERING OPERATIONS MANAGER. Air National Guard, Kulis Air Field, Anchorage, AK (1997-00).
Supervisor: Lt. Colonel Patrick Noble (444) 555-6666
Pay grade: Major (O-4) **Hours worked per week:** 40+
Duties: Provided supervisory oversight and training to 44 engineering personnel. Interviewed and hired all employees. Managed the programming, design, and construction of 17 projects totaling more than $22 million, including a complex, environmentally sensitive $1.1 million project to remove and replace underground fuel storage tanks. Reviewed and evaluated cost proposals and designs submitted by architectural and engineering firms. Conducted negotiations with officials from those companies to arrive at a more favorable price and

obtain required modifications to their design specifications. Identified and programmed required maintenance and repair for a 10,000 acre, 329 facility physical plant comprising more than 460,000 square feet.

- Effectively executed critical fast-track construction when facilities were not in use, completing all necessary projects during the winter and keeping construction on schedule in spite of adverse building conditions and temperatures as low as -30° Fahrenheit.
- Projects completed included a 9,000-foot runway repair, an addition to and replacement of the refueler haul road, an $8 million aircraft ramp refurbishment, as well as construction of a $2.5 million Radar Control facility, design and construction of a $2.7 million Communications facility, and construction of a $2.3 million maintenance facility.

PROGRAM MANAGER. United States Air Force, Eielson AFB, AK (1995-97).
Supervisor: Major Dustin Cuny **Pay grade:** Captain (O-3)
Hours worked per week: 40+
Duties: Oversaw all phases of technical management for a unique five-year, $150 million contract providing quick-reaction critical support to Space Systems Division launches, launch-related facilities, and satellite operations and tracking facilities. Coordinated and directed complex planning and funding from three separate operating locations using five different appropriations channeled through 15 different using organizations. Expertly managed the transition to a new contractor from an organization that held this contract for 26 years, ensuring continuous support to space systems ranging from meteorological, surveillance, and global positioning satellites to Atlas and Titan launch vehicles with no launch delays.

- Reviewed and evaluated a proposal for a fast-track repair of Delta Launch PA; approved the project at a cost of only $778,000, saving the government more than $1 million dollars from the contractor's original cost projection while ensuring the project was completed in time to meet NASA's launch window.
- Developed and implemented a new Total Quality Management (TQM)-based metric measurement system designed to clarify critical issues to upper management and eliminate bureaucratic delays in the processing system for putting work on the contract.

MECHANICAL ENGINEER. United States Air Force, Altus AFB, OK (1990-95).
Supervisor: Mr. David Dane **Pay grade:** 2LT (O-1)
Hours worked per week: 40+
Duties: Provided Civil Engineering support and technical assistance, including but not limited to preparation of construction drawings, specifications, and estimation of material and labor costs. Designed plumbing, electrical, HVAC, and other systems, providing a variety of professional engineering services required in programming and in the design for construction, modification, and maintenance of facilities.

- Provided technical assistance during the negotiation phase and follow-on engineering assistance in the design phase of more than $12.1 million in construction projects that were bid on by outside architectural and engineering firms.
- Recognized for my "superb technical input to the Commander's Energy Initiative."

EDUCATION & TRAINING

Master of Science in Mechanical Engineering, University of Arizona, Tucson, AZ, 1996.
Bachelor of Science in Mechanical Engineering, Arizona State University, Tempe, AZ, 1989.
Completed numerous leadership and management development courses sponsored by the United States Air Force, including:

- **Air Command & Staff College**, October 2002.
- **Squadron Officer School** – Distinguished Graduate, December 1995; member of the squadron Right of Line flight for top performance in academic and leadership competition.
- **Officers Training School**, May 1990.

FEDERAL RESUME OR RESUMIX

BRENT MORALES
SSN: 123-45-6789
1110½ Hay Street
Fayetteville, NC 28305
Phone: (910) 483-6611
Email: preppub@aol.com

ANNOUNCEMENT NUMBER: 05MAR4836611

ELECTRICAL POWER SAFETY SUPERVISOR

EXPERIENCE SUMMARY

Extensive knowledge of total quality management methodology, tools, and techniques. Knowledge of directives pertaining to U.S. Air Force disaster preparedness programs and activities to project, plan, and manage disaster preparedness programs. Knowledge of Air Base Operability program, governing directives and methods and procedures for program implementation. Extensive technical and supervisory experience related to all aspects of the repair, maintenance, and operation of power plant equipment in accordance with manufacturer's specifications, technical orders, and shop operating procedures. Highly resourceful and technically expert troubleshooter and problem solver in testing and emergency situations. Extensive background in personnel supervision, industrial maintenance, preventive maintenance, and electrical power production.

EXPERIENCE

ELECTRICAL POWER SAFETY SUPERVISOR. U.S. Air Force Reserve, Tinker AFB, OK (Jul 2002 – Mar 2005). 480 hours per year, Supervisor's name and phone: Charles Knowles, (910) 483-6611.
Supervise, inspect, maintain, operate, and repair mobile, gas, and diesel power generators ranging in size from 5kw to 750kw. Maintain all associated equipment such as portable load banks. Utilize power tools and precision measurement equipment. From 17 Jul 2002-30 Jul 2003, deployed to the 8th Air Expeditionary Group (AEG) in support of Operation Iraqi Freedom at Baghdad, Iraq. Maintained over $6 million worth of electrical power generation and ground support equipment. Was Key Control Operator for a 6 megawatt, 4160 volt, prime power plant consisting of eight Mobile Electric Power-012A (MEP) generators supporting 95% of the base electrical requirements. Professionally performed several 300-hour Preventive Maintenance Inspections (PMIs) on various MEP series and commercial generator sets while also providing well-maintained generators for reliable electrical power. Performed daily inspections on 11 generators providing prime electrical power to remote locations. Was evaluated as "an outstanding performer and a tremendous asset to the civil engineers at Baghdad. Was commended for "superior technical knowledge which allowed training of less qualified technicians on mission-critical emergency generators. Became known for my strong troubleshooting and problem-solving skills: on numerous occasions, performed rapid maintenance actions on broken generator engineers. Install and operate electrical power plants, mobile generators, distribution equipment, automatic transfer panel, airfield lighting, and aircraft arresting systems.

ELECTRICAL POWER PRODUCTION MECHANIC. U.S. Air Force Reserves, Kirtland AFB, NM (Oct 2000 - Jun 2002). 480 hours per year, Supervisor's name and phone: Martin Smith, (910) 483-6611. Inspected, maintained, operated, and repaired mobile, gas, and diesel power generators ranging in size from 5kw to 100kw and Mobile BAK-12 Aircraft Arresting Barriers. Maintained all associated equipment such as air compressors, pumps, and portable load banks. Utilized power tools and precision measurement equipment. On a formal performance evaluation, was praised in these words: "SSG Morales performs all assigned duties in a professional manner and sets the standard for others to follow. Was described as "a multi-skilled individual capable of handling and completing any type of task."

SHOP SAFETY MONITOR. U.S. Air Force, Charleston AFB, SC (Jan 1990 - Jun 2000). 40 hours per week, Supervisor's name and phone: Aiden Rainey, phone unknown. Was promoted to supervise six people as supervisor for the Main Engine Rebuilt Section, Main Component Rebuilt Section and the Cylinder Head Rebuild Section. Supervised 18 people in the inspection, maintenance, and repair of power generators and support equipment for all military bases in southeastern U.S. Performed additional duty as Safety Monitor. Earned respect for my ability to perform inspection and repair activities to the depot level overhaul and maintenance of both prime and standby power generating units at 15 remote installations throughout South Carolina. Disassembled, cleaned, visually inspected, repaired, assembled, and tested diesel electric generating systems, air compressors, aircraft arresting barriers, and all mechanical components utilizing the latest test equipment and procedures. Repaired, modified, and overhauls power plant equipment in accordance with manufacturer's specifications, technical orders, and current shop procedures. Was evaluated on a formal performance evaluation as "a highly motivated and dedicated NCO who continuously produces excellent results from his every tasking." Was praised for my outstanding performance in correcting numerous mechanical problems in order to keep generators in service. Was praised in a formal performance evaluation for "initiative and versatility" while working as part of team at Concord Airport repairing the prime power plant's commercial buss breakers. The repairs enabled the base to transfer the commercial power source when available without interference to base missions. During one assignment as manager of the Small Engine Repair Section, implemented procedures which improved productivity and efficiency. During one special assignment which involved correcting a malfunction on a Caterpillar D353 generator, utilized my expertise to design, manufacture, and install a fuel line system, which restored the unit to fully operational status. Served as Safety Monitor and became known for my aggressive emphasis on safety and quality assurance at all times.

EDUCATION

Associate of Applied Science Degree in Automotive/Diesel Technology, Albuquerque Technical Vocational Institute, Albuquerque, NM, Jan 2002.
Master Mechanic Course, Albuquerque Technical Vocational Institute, Albuquerque, NM, Jan 2002.

TRAINING

Completed training in Electrical Standards, U.S. Department of Labor, Occupational Health and Safety Administration, Washington, D.C., Jan 2005.
Onan Generator Qualification Certification Course leading to Onan Generator Qualification, Oklahoma City, OK; Jan 2003.
Hazardous Material Pallet Loading Course, U.S. Air Force, May 2002. Total Quality Management, U.S. Air Force, Albuquerque, NM, Jul 2001. Diesel Engine Overhaul and Generator Set Operation Course, U.S. Air Force, Tinker AFB, OK, Dec 1997. Non-Commissioned Officer Leadership School, U.S. Air Force, Charleston AFB, SC, Oct 1992. Electrical Power Production Technician Course, U.S. Air Force, Pope AFB, NC, Oct 1986.

LICENSES & CERTIFICATES

Certified as a Level II Onan Generator Mechanic, Jan 2003.
Certified Service Technician, Jan 2003.
Mobile Air Conditioning Society License, August 2001.

FEDERAL RESUME OR RESUMIX

VICTOR A. NEILSON
1110 ½ Hay Street
Fayetteville, NC 28305
Home: 910-483-6611
Cell: 910-483-6611
E-mail: preppub@aol.com

SSN: 123-45-6789
Date of birth: 01/01/1973
Country of Citizenship: United States
Veteran's Preference: 5%

QUALITY ASSURANCE REPRESENTATIVE

EXPERIENCE

May 2005-present: **QUALITY ASSURANCE REPRESENTATIVE.** PRP, Inc., 1110½ Hay Street, Dallas, TX 28305. $48,000 per annum. Supervisor: Corey Bailey, 910-483-6611. Am a key member of the management team of a company which leases critical industrial equipment including power generators, temperature control equipment, and air compressors to customer organization which include chemical plants, manufacturing plants, building contractors, military organizations, and other customers. Perform major functions including the following:

- **Quality Assurance:** Perform analysis/investigations to ensure proper maintenance of power generators, temperature control equipment, and air compressors. Assure that leased equipment is in quality operating condition prior to delivery to customers. When leased equipment is returned, check material for evidence of carelessness or misuse of equipment or property. Determine liability for property that is damaged, lost, or destroyed. Conduct and document safety and environmental inspections. Coordinate all labor for normal or emergency repair of equipment.
- **Customer Service:** Coordinate with customers. Manage delivery schedules of leased equipment.
- **Logistics Management:** Coordinate the logistics involved in providing customers with leased equipment. Organize transportation of equipment via commercial carriers and expertly prepare and review all paperwork and documentation to assure completeness for future audits.
- **Contract Negotiation:** Negotiate contracts with customers for leased equipment. Negotiate key details of leases including price, delivery, customer support services, and other issues.
- **Experience with Military Contracting:** Coordinate with military contracting representatives and have become knowledgeable of the process of responding to solicitations as well as providing quality assurance information to contract specialists.

Oct 2003-Apr 2005: **CUSTOMER SERVICE REPRESENTATIVE & CONTRACT SPECIALIST.** United Rentals, 1110½ Hay Street, Dallas, TX 28305. $40,000 per annum. Supervisor: Chris Turner, Phone 910-483-6611. Prospected for new accounts while servicing existing customers. Coordinated with all levels of management in various industries including manufacturing and construction. Negotiated long-term and short-term leases for major pieces of construction equipment. Was handpicked to train new customer service representatives.

Aug 2001-Oct 2003: **EQUIPMENT COORDINATOR & CONTRACT SPECIALIST.** John Equipment, 1110½ Hay Street, Dallas, TX 28305. $30,000. Supervisor: Richard Stone, Phone 910-483-6611. For a $6 million fleet of construction equipment, prepared and maintained rental/lease agreements and handled collections as needed. Edited and filed all rental agreement reports, customer reports, and equipment reports for regular business day operations.

Dec 1999-Jul 2001: **ENGINEERING WORK-STUDY.** Smith Memorial Hospital, 1110½ Hay Street, Dallas, TX 28305. Supervisor: Al Morris, Phone 910-483-6611. While completing requirements for my Bachelor of Business Administration degree, excelled in a work-study program with the Smith Memorial Hospital during which I repaired and maintained all mechanical hospital equipment. Worked without supervision while repairing and maintaining ceiling tile, sinks, and drains throughout the hospital. Also worked in the X-ray room, where I prepared and maintained records for customers and agency use while interacting with all levels of hospital personnel.

Sept 1997-Nov 1999: **POWER GENERATION EQUIPMENT REPAIR SUPERVISOR.** Dallas Army National Guard, Detachment, 125th Engineer BN, 1110½ Hay Street, Dallas, TX 28305. Rank: E-5. Supervisor: Chief Warrant Officer Caleb Lonestar, Phone 910-483-6611. As a proud member of the National Guard, maintained and repaired diesel generator systems and vehicles throughout the unit. Handled extensive responsibility for quality assurance as I inspected and repaired all equipment in order to assure 100% serviceability. Became highly experienced in inspecting and operating all types of vehicles ranging from quarter-ton trucks, to five-ton cargo trucks, to forklifts. Trained, supervised, and counseled junior employees in equipment operation and military matters. As a supervisor, was responsible for maintaining safety in all phases of maintenance and equipment movement.

Sept 1990-Sept 1997: **GENERATOR REPAIR SPECIALIST & QUALITY ASSURANCE TECHNICIAN.** U.S. Army, Fort Hood, TX 28305. Rank E-4. Supervisors: Multiple. Phone unknown. On active duty with the U.S. Army, maintained 40 generator systems while continuously performing quality assurance inspections. Inspected paperwork to ensure that correct repair and maintenance procedures were followed. Trained and supervised junior employees, and assigned specific duties as well as tools and equipment to employees.

COMPUTER EXPERTISE

Experienced in utilizing custom software programs for procurement, supply management, logistics management, and accounting management.
Skilled in utilizing Microsoft Word and the MS Office including Excel and PowerPoint.
Completed computer training in Windows and Word.

EDUCATION & TRAINING

Completed the Bachelor of Business Administration (B.B.A.) degree, Southern Methodist University, Dallas, TX, 2000.
Graduated from Primary Leadership Development Course, Reserve Component Noncommissioned Officers Course, Fort Hood, 1997.
Graduated from the Ordnance Center and School, Power Generation Equipment Repairer Course, Fort Bragg, NC, 1993. Completed training in these and other areas:

Arc Welder Repair	Gas Engine Repair	Circuit Board Repair
Exciter and Exciter Regulator Repair	Ignition Systems	Starter Motor Repair
Liquid Cooling	Fuel Circuits	Battery Charging

MEDALS & HONORS

Certificates of Achievement; Army Commendation Medal; numerous letters of commendation and appreciation.

FEDERAL RESUME OR RESUMIX

JOEL DUSTIN SPENCER
SSN: 000-00-0000
1110 1/2 Hay Street
Fayetteville, NC 28305
Home: 910-483-6611
E-mail: preppub@aol.com

Job title in announcement: SAFETY INSPECTOR
Grade(s) applied for: FG—1825-9/11
Announcement number: FAA-ASI-11-111-11111A

SAFETY INSPECTOR

EXPERIENCE

SAFETY INSPECTOR. 11th Wing, Bolling AFB, DC (08/92-present). Supervisor: F. Sweeney, Bolling AFB, D.C., 910-483-6611. $22.15 per hour, 40 hours per week.

Quality assurance and inspection duties: Inspect entire aircraft and aircraft systems, surfaces, and assemblies for serviceability and proper operation. Complete a four-phased inspection process comprised of:

- **Preinspection phase:** Tow aircraft to wash areas and inspect aircraft for cleanliness and corrosion control. Perform engine run up and aircraft systems checkout and defueling. Jack and level aircraft to ensure exact positioning of aircraft during inspection process for weight, balance, alignment checks, and functional alignment of systems. Open up aircraft by removing panels, leading edges, and other structural surfaces.
- **Look phase and fix phases:** Perform visual, dimensional, and operational inspections of aircraft operating or control systems, surfaces, assemblies and related parts, accessories, or components. Diagnose and correct malfunctions. Inspect entire aircraft for cracks, corrosion, loose or missing nuts, bolts, rivets, or other aircraft hardware. Inspect electrical wiring and hydraulic tubing and hoses for serviceability, and replace wheels and tires as well as brakes and struts as necessary. Remove and replace leading edges for repair or to facilitate maintenance. Identify defects and take corrective action. Inspect, operate, troubleshoot, repair, replace, and modify flight surface structures such as the fuselage, wings, and empennage **flight control systems** such as ailerons, elevators, and rudders; **mechanical systems** to include rigging; **duel rail systems**, and life raft **ejection systems; hydraulic, fuel, and pressurization systems; environmental systems;** wheel and landing gear assemblies and other aircraft systems, assemblies, and surfaces. Repair, rework, and replace parts or components of systems to include proper system operation and alignment with other systems. Rig, adjust, and synchronize controls cables, linkages and linkage mechanisms, actuating, and other mechanisms to ensure proper tension and operation.
- **Postinspection phase:** Conduct postinspection phase involving transfer of the aircraft from the inspection hangar to the postinspection area. Handle engine run up and activation of systems for completion of the maintenance operation check (MOC). Troubleshoot and repair malfunctioning general aircraft systems and assist shop specialists on specific/single aircraft systems.

AIRCRAFT MECHANIC & REPAIR AND RECLAMATION TECHNICIAN. 33rd Fighter Wing, Elgin AFB, UT (06/1987-09/1992). Supervisor: C. Williams, 910-483-6611. $13.08 per hour, 40 hours per week. Supervised between 2-4 employees while expertly performing quality assurance duties. Tested and observed the operation of aircraft flight control systems and surfaces to determine serviceability and need for replacement or repair. Checked cable tension, synchronization of cockpit control with control surfaces, as well as the condition of items including pull-push rods, bell cranks, ailerons, rudder, flaps, and elevators. Disassembled, removed, repaired, replaced, and reassembled control systems and components. Aligned control systems and tested control systems and surfaces after making repairs and replacements. Built up and tore down aircraft wheel assemblies by removing the wheel and inner tube from tire. Examined wheel bearings, tubes, and rims for defects and signs of wear. Cleaned and packed bears, reassembled wheel assemblies, and checked tires for air pressure and leaks.

Performed periodic inspections of two main landing gears, nose gear, and nose gear steering. Disassembled, reassembled, and performed landing gear retraction testing. Utilized portable hydraulic test stand to diagnose malfunctions. Made adjustments; repaired and replaced parts to correct any malfunction's. Repaired reclaimed parts for turn-in to supply.

- Performed as a fully qualified and certified technician to accomplish repairs in the cockpit "white area."

Highlights of other experience:

CO-OWNER, FAMILY RESTAURANT. Family Experience Restaurant, Willford Plaza, Thibodaux, LA (06/1982-08/1986). Self-employed. $15.00 and up, 40 hours per week. Hired, trained, and managed dozens of employees while supervising all operations of a family-owned restaurant. Scheduled employees, ordered food items, handled inventory control, and provided outstanding customer relations.

QUALITY CONTROL INSPECTOR. Louisiana Manufacturing, Thibodaux, LA (08/1983-06/1985). Supervisor: L. Cruces, phone unknown. $10.48 hour, 40 hours per week. Performed quality control inspections of systems and manufactured products for industrial conveyor pulleys and drive belts.

EDUCATION & TRAINING

Thibodaux Sr. High School, Thibodaux, LA , 1973
Airframe & Power Plant Certificate, May 2003; Student Pilot (65 hours flight time); USAFR Leadership Development Training, 1994; Aircraft Battle Damage Repair, 1992; CMD NCO Academy, 1989; Air Force Documentation Maintenance, 1988; Total Quality Management Training, 1997; Aircraft Maintenance School, 1977-78; Mortar School, 1973; NCO Orientation Course, 1980; OJT Orientation, 1986

MEDALS & AWARDS

Performance Award, Sept 1998, Sept 1997, Sept 1994, April 1992, April 1981; Sustained Superior Performance Award, 1989; Notable Achievement Award, April 1988.

DESIGN SKILLS

Designed a new tool for the KC-135

SECURITY CLEARANCE

Can pass the most rigorous background investigation

ERIC LANDON HINKLE
111-11-1111
FOOD SAFETY INSPECTOR (PROCESSED PRODUCTS), GS-483-6611 (SJ-483)
Announcement Number PREP-PB-483-6611

FOOD SAFETY INSPECTOR (PROCESSED PRODUCTS), GS-483-6611 (SJ-483) Announcement # PREP-PB-483-6611
Element (A)
Page One

ELEMENT (A): Ability to analyze findings and develop sound recommendations for safety and quality improvement.

With regard to the laboratory, I make nightly checks on the Quality Assurance department to evaluate their records. At any time that we are suspicious that there could be a problem of any type, I organize my workload to provide adequate coverage so that I can make myself available to consult with plant management on resolving the potential situation. For example, the Blakely plant had a history of failing moisture tests, therefore I discussed the problem with plant management, who determined that they would run two moisture tests per day to ensure that moisture tests were within compliant levels. I then provided oversight and monitored the records and results of these tests until both myself and plant management were confident that moisture levels were in compliance and the problem had been successfully addressed.

In an earlier situation, during an inspection of the Early Farms, Inc. plant in Blakely, GA, I observed a condensation problem that could have led to possible product adulteration, as the condensation was occurring directly over the product contact zones. I consulted with the processing superintendent and together we brainstormed possible ways to remedy the problem. I facilitated the problem-solving process, with the end result that plant personnel were able to solve their own problem internally.

While conducting a Pre-Operational Sanitation Inspection, I discovered a situation that could have led to a suspension, involving "black specks" showing up on the conveyor belt. During the course of the inspection, I identified the problem with black specks on various conveyors. I met with the Inspector in Charge (IIC), the Pre-Op Superintendent, the Front-end Plant Manager, and the Back-end Plant Manager, in order to discuss orally the severity of the problem. I explained that non-correction of the problem could lead to possible suspension, and I provided the leadership during this verbal exchange which allowed the plant to quickly identify solutions to this problem. Within a matter of weeks the problem was completely corrected, and that problem has never reoccurred.

On another occasion at the Blakely plant, I discovered a possible product adulteration situation involving System 1 and System 2 post chillers, namely that there was a product-on-floor problem. This could have led to product adulteration, since there were as many as 40-50 chickens on the floor two or three times per night. At the weekly Performance Based Inspection System (PBIS) meeting, I voiced my concerns to personnel from plant management, quality assurance, hazard coordination, and others. By facilitating a discussion among plant personnel regarding possible methods of addressing and correcting this problem, I ensured that the appropriate individuals at the meeting were aware of the problem and its severity; they, in turn, communicated verbally with plant operations personnel, and I soon observed that the plant had taken appropriate corrective and preventive action to resolve this problem.

Education and Training Related to this Element:
Currently attending Albany State University majoring in Mathematics, 2002-present.
Majored in math at Columbus State University, 1995-1998.
Have completed the following training courses related to inspection and processing:

- Basic Processed Products and Quality Control Inspection 501, United States Department of Agriculture, Food Safety and Inspection Service Human Resources Development Division, Albany State University, GA (2004).
- Field Automation and Information Management training, United States Department of Agriculture, Food Safety and Inspection Service Human Resources Development Division, Albany State University, GA (2004).
- HACCP training, Columbus Livestock Building, Columbus State University, GA (2003).
- Pre-HACCP training, Columbus State University, GA (2002).
- PBIS 03 training, Columbus State University, GA (2002).
- 705C On-line Slaughter Quality Control, United States Department of Agriculture, Food Safety and Inspection Service Human Resources Development Division, Albany State University, GA (2001).
- Livestock Slaughter Inspection Cross-Training 305X, United States Department of Agriculture, Food Safety and Inspection Service Human Resource Development Division, Albany State University, GA (2000).
- Basic Poultry Inspection 703C, United States Department of Agriculture, Food Safety and Inspection Service Human Resource Development Division, Albany State University, GA (1998).
- Attended a two-day seminar along with the Sanitation Manager and Inspector-In-Charge to implement a pre-op program; designed a chart to determine effective sanitation inspection procedures, Blakely, GA (1998).

Gained valuable knowledge related to viruses, bacteria, and parasites, as well as effective preventive measures while participating in various science classes including Biology and Chemistry at Albany High School, Albany, GA (1995), and Columbus State University, GA (1995-98).

FOOD SAFETY INSPECTOR (PROCESSED PRODUCTS), GS-483-6611 (SJ-483) Announcement # PREP-PB-483-6611
Element (A)
Page Two

EXAMPLE OF A KSA

DEXTER LEMON
111-11-1111
SAFETY SPECIALIST, GS-0260
Announcement #AC-01-000

SAFETY SPECIALIST, GS-0260
Announcement #AC-01-000
KSA #1

KSA #1: Knowledge of Safety and Occupational Health.

Throughout a military career culminating in the rank of Sergeant Major, became known as professional who emphasized the importance of safety and occupational health. Since 2002, I have been actively focused on managing EEO programs and ensuring large military organizations operate active full-time programs for these vital issues.

Related experience:
As the Equal Opportunity Program Manager since 5/05, I have been excelling in a position normally held by a LTC. I develop, direct, evaluate, and monitor a comprehensive program for equal opportunity, affirmative action, and complaint adjudication for a military community which includes more than 20,000 soldiers as well as family members and Department of Defense (DOD) civilians. As the 82d Airborne Division's principal advisor and "subject matter expert" on these issues, I provide input into the formulation of the installation's personnel policies, promotion of EO awareness, and implementation of a sexual harassment prevention program. I monitor and assess implementation of EO policy by contractors and conduct impact studies on minorities, women, and handicapped personnel for an affirmative action program primarily concerned with improving minority and female representation in administrative and technical positions. Analyze management practices, organizational structure, employment patterns, and lines of progression to determine the impact on equal opportunity and upward mobility. Prepare an annual affirmative action plan and assure its compliance with EO principles. Make personal appearances and presentations to community organizations, service personnel, schools and universities, and other public and private groups; use these occasions to exchange ideas and gain support. Train and educate managers, supervisors, and employees. Have developed a series of training packages used to teach classes in the full range of related subject matter. Counsel personnel who feel they have been the object of discrimination; ensure complaints are responded to impartially, quickly, and fairly; have assisted in more than 100 complaints over the last four years.

From 1/02-4/05, served as Equal Opportunity Advisor and Installation EO NCOIC for Fort Bragg, NC. Became known as an expert on equal opportunity, sexual harassment, and discrimination issues and excelled in guiding the programs of a community of 19,000 soldiers as well as family members and civilian employees. Played a major role in planning and conducting numerous Heritage Month Observances which emphasized multicultural awareness and involvement.

Education and training:
Am pursuing a **B.S. degree Human Relations and Human Resources,** Fayetteville State University, Fayetteville, NC; degree to be awarded August 2006.
Completed extensive U.S. Army management and leadership training including Defense Equal Opportunity Management Institute, 2005, certification in Mediation/ADR.
Defense Equal Opportunity Management Institute, 2001, concentrated in Human Relations and Race Relations was certified as an Equal Opportunity Advisor and Manager Carolina Central Institute, 2005, studied Real Estate and Fair Housing Laws and was certified as a Real Estate Agent and Fair Housing Monitor in North Carolina.

DEXTER LEMON
111-11-1111
SAFETY SPECIALIST, GS-0260
Announcement #AC-01-000

KSA #2: Skill in written communication.

SAFETY SPECIALIST, GS-0260 Announcement #AC-01-000 KSA #2

Overview of skill in written communication:
Throughout a military career culminating in my present rank of Master Sergeant, I have been known as an articulate professional with a talent for developing informative and interesting information geared to the specific audience. My ability to develop effective, informative, and easily understood written material for all types of audiences has been an important factor in my success in EEO program development and management.

Experience:
As the Equal Opportunity Program Manager since 5/05, am excelling in a position normally held by a LTC. I apply my written communication skills while involved in planning, directing, implementing, evaluating, and monitoring a comprehensive program for equal opportunity, affirmative action, and complaint adjudication program for a military community which includes more than 20,000 soldiers as well as family members and Department of Defense (DOD) civilians. Serve as principal advisor and "subject matter expert" on these issues to include providing written and verbal input into the formulation of the installation's personnel policies, promotion of EO awareness, and implementation of a sexual harassment prevention program. Prepare written reports based on my analysis of management practices, organizational structure, employment patterns, and lines of progression and their impact on equal opportunity and upward mobility. Have developed a series of training packages used to teach classes in the full range of related subject matter. Totally revamped the EO program and developed numerous publications emphasizing race relations, extremism, sexual harassment, and policy implementation.

As the Equal Opportunity Program Advisor and Installation EO NCOIC for Fort Bragg, NC (1/02-4/05), became known as an expert on equal opportunity, sexual harassment, and discrimination issues and excelled in guiding the programs of a community of 19,000 soldiers as well as family members and civilian employees. Applied written communication skills while developing materials used as the Primary Instructor for an 80-hour EO Leader's Course; trained more than 210 subordinate unit EO Representatives and coordinated quarterly seminars. Was active in investigating several reports of alleged discrimination, all of which were fairly and impartially resolved, and in creating concise and easily understood reports of each case.

Education and training:
Am pursuing a B.S. degree in **Human Relations and Human Resources,** Fayetteville State University, Fayetteville, NC; degree to be awarded August 2006.
Completed extensive U.S. Army management and leadership training including:
Defense Equal Opportunity Management Institute, 2005, earned certification in Mediation/ADR.
Defense Equal Opportunity Management Institute, 2001, concentrated in Human Relations and Race Relations was certified as an Equal Opportunity Advisor and Manager
Carolina Central Institute, 2005, studied Real Estate and Fair Housing Laws and was certified as a Real Estate Agent and Fair Housing Monitor in North Carolina.

ABOUT THE EDITOR

Anne McKinney holds an MBA from the Harvard Business School and a BA in English from the University of North Carolina at Chapel Hill. A noted public speaker, writer, and teacher, she is the senior editor for PREP's business and career imprint, which bears her name. Early titles in the Anne McKinney Career Series (now called the Real-Resumes Series) published by PREP include: *Resumes and Cover Letters That Have Worked, Resumes and Cover Letters That Have Worked for Military Professionals, Government Job Applications and Federal Resumes, Cover Letters That Blow Doors Open,* and *Letters for Special Situations.* Her career titles and how-to resume-and-cover-letter books are based on the expertise she has acquired in 25 years of working with job hunters. Her valuable career insights have appeared in publications of the "Wall Street Journal" and other prominent newspapers and magazines.

PREP Publishing Order Form You may purchase our titles from your favorite bookseller! Or send a check, money order or your credit card number for the total amount*, plus $4.00 postage and handling, to PREP, 1110 1/2 Hay Street, Fayetteville, NC 28305. You may also order our titles on our website at www.prep-pub.com and feel free to e-mail us at preppub@aol.com or call 910-483-6611 with your questions or concerns.

Name: ____

Address: ____

E-mail address: ____

Payment Type: ☐ Check/Money Order ☐ Visa ☐ MasterCard

Credit Card Number: ______________________ Expiration Date: ________

Put a check beside the items you are ordering:

☐ $16.95—REAL-RESUMES FOR RESTAURANT, FOOD SERVICE & HOTEL JOBS. Anne McKinney, Editor
☐ $16.95—REAL-RESUMES FOR MEDIA, NEWSPAPER, BROADCASTING & PUBLIC AFFAIRS JOBS. Anne McKinney, Editor
☐ $16.95—REAL-RESUMES FOR RETAILING, MODELING, FASHION & BEAUTY JOBS. Anne McKinney, Editor
☐ $16.95—REAL-RESUMES FOR HUMAN RESOURCES & PERSONNEL JOBS. Anne McKinney, Editor
☐ $16.95—REAL-RESUMES FOR MANUFACTURING JOBS. Anne McKinney, Editor
☐ $16.95—REAL-RESUMES FOR AVIATION & TRAVEL JOBS. Anne McKinney, Editor
☐ $16.95—REAL-RESUMES FOR POLICE, LAW ENFORCEMENT & SECURITY JOBS. Anne McKinney, Editor
☐ $16.95—REAL-RESUMES FOR SOCIAL WORK & COUNSELING JOBS. Anne McKinney, Editor
☐ $16.95—REAL-RESUMES FOR CONSTRUCTION JOBS. Anne McKinney, Editor
☐ $16.95—REAL-RESUMES FOR FINANCIAL JOBS. Anne McKinney, Editor
☐ $16.95—REAL-RESUMES FOR COMPUTER JOBS. Anne McKinney, Editor
☐ $16.95—REAL-RESUMES FOR MEDICAL JOBS. Anne McKinney, Editor
☐ $16.95—REAL-RESUMES FOR TEACHERS. Anne McKinney, Editor
☐ $16.95—REAL-RESUMES FOR CAREER CHANGERS. Anne McKinney, Editor
☐ $16.95—REAL-RESUMES FOR STUDENTS. Anne McKinney, Editor
☐ $16.95—REAL-RESUMES FOR SALES. Anne McKinney, Editor
☐ $16.95—REAL ESSAYS FOR COLLEGE AND GRAD SCHOOL. Anne McKinney, Editor
☐ $25.00—RESUMES AND COVER LETTERS THAT HAVE WORKED. McKinney, Editor
☐ $25.00—RESUMES AND COVER LETTERS THAT HAVE WORKED FOR MILITARY PROFESSIONALS. McKinney, Editor
☐ $25.00—RESUMES AND COVER LETTERS FOR MANAGERS. McKinney, Editor
☐ $25.00—GOVERNMENT JOB APPLICATIONS AND FEDERAL RESUMES. McKinney, Editor
☐ $25.00—COVER LETTERS THAT BLOW DOORS OPEN. McKinney, Editor
☐ $25.00—LETTERS FOR SPECIAL SITUATIONS. McKinney, Editor
☐ $16.95—REAL-RESUMES FOR NURSING JOBS. McKinney, Editor
☐ $16.95—REAL-RESUMES FOR AUTO INDUSTRY JOBS. McKinney, Editor
☐ $24.95—REAL KSAS--KNOWLEDGE, SKILLS & ABILITIES--FOR GOVERNMENT JOBS. McKinney, Editor
☐ $24.95—REAL RESUMIX AND OTHER RESUMES FOR FEDERAL GOVERNMENT JOBS. McKinney, Editor
☐ $24.95—REAL BUSINESS PLANS AND MARKETING TOOLS ... Samples to use in your business. McKinney, Editor
☐ $16.95—REAL-RESUMES FOR ADMINISTRATIVE SUPPORT, OFFICE & SECRETARIAL JOBS. Anne McKinney, Editor
☐ $16.95—REAL-RESUMES FOR FIREFIGHTING JOBS. Anne McKinney, Editor
☐ $16.95—REAL-RESUMES FOR JOBS IN NONPROFIT ORGANIZATIONS. Anne McKinney, Editor
☐ $16.95—REAL-RESUMES FOR SPORTS INDUSTRY JOBS. Anne McKinney, Editor
☐ $16.95—REAL-RESUMES FOR LEGAL & PARALEGAL JOBS. Anne McKinney, Editor
☐ $16.95—REAL-RESUMES FOR ENGINEERING JOBS. Anne McKinney, Editor
☐ $22.95—REAL-RESUMES FOR U.S. POSTAL SERVICE JOBS. Anne McKinney, Editor
☐ $16.95—REAL-RESUMES FOR CUSTOMER SERVICE JOBS. Anne McKinney, Editor
☐ $16.95—REAL-RESUMES FOR SAFETY & QUALITY ASSURANCE JOBS. Anne McKinney, Editor

__________ **TOTAL ORDERED**

__________ **(add $4.00 for shipping and handling)**

__________ **TOTAL INCLUDING SHIPPING *PREP** *offers volume discounts. Call (910) 483-6611.*

THE MISSION OF PREP PUBLISHING IS TO PUBLISH BOOKS AND OTHER PRODUCTS WHICH ENRICH PEOPLE'S LIVES AND HELP THEM OPTIMIZE THE HUMAN EXPERIENCE. OUR STRONGEST LINES ARE OUR JUDEO-CHRISTIAN ETHICS SERIES AND OUR REAL-RESUMES SERIES.